Marine Mammals and Low-Frequency Sound
Progress Since 1994

Committee to Review Results of ATOC's
Marine Mammal Research Program

Ocean Studies Board

Commission on Geosciences, Environment, and Resources

National Research Council

NATIONAL ACADEMY PRESS
Washington, D.C.

NATIONAL ACADEMY PRESS • 2101 Constitution Avenue, NW • Washington, DC 20418

NOTICE: The project that is the subject of this report was approved by the Governing Board of the National Research Council, whose members are drawn from the councils of the National Academy of Sciences, the National Academy of Engineering, and the Institute of Medicine. The members of the committee responsible for the report were chosen for their special competencies and with regard for appropriate balance.

This report was supported by grants from the Defense Advanced Research Projects Agency and the Office of Naval Research. The views expressed herein are those of the authors and do not necessarily reflect the views of the sponsors.

Library of Congress Catalog Number 00-101832
International Standard Book Number 0-309-06886-X

Additional copies of this report are available from:
National Academy Press
2101 Constitution Avenue, NW
Box 285
Washington, DC 20055
800-624-6242
202-334-3313 (in the Washington metropolitan area)
http://www.nap.edu

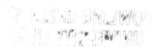

THE NATIONAL ACADEMIES

National Academy of Sciences
National Academy of Engineering
Institute of Medicine
National Research Council

The **National Academy of Sciences** is a private, nonprofit, self-perpetuating society of distinguished scholars engaged in scientific and engineering research, dedicated to the furtherance of science and technology and to their use for the general welfare. Upon the authority of the charter granted to it by the Congress in 1863, the Academy has a mandate that requires it to advise the federal government on scientific and technical matters. Dr. Bruce M. Alberts is president of the National Academy of Sciences.

The **National Academy of Engineering** was established in 1964, under the charter of the National Academy of Sciences, as a parallel organization of outstanding engineers. It is autonomous in its administration and in the selection of its members, sharing with the National Academy of Sciences the responsibility for advising the federal government. The National Academy of Engineering also sponsors engineering programs aimed at meeting national needs, encourages education and research, and recognizes the superior achievements of engineers. Dr. William A. Wulf is president of the National Academy of Engineering.

The **Institute of Medicine** was established in 1970 by the National Academy of Sciences to secure the services of eminent members of appropriate professions in the examination of policy matters pertaining to the health of the public. The Institute acts under the responsibility given to the National Academy of Sciences by its congressional charter to be an adviser to the federal government and, upon its own initiative, to identify issues of medical care, research, and education. Dr. Kenneth I. Shine is president of the Institute of Medicine.

The **National Research Council** was organized by the National Academy of Sciences in 1916 to associate the broad community of science and technology with the Academy's purposes of furthering knowledge and advising the federal government. Functioning in accordance with general policies determined by the Academy, the Council has become the principal operating agency of both the National Academy of Sciences and the National Academy of Engineering in providing services to the government, the public, and the scientific and engineering communities. The Council is administered jointly by both Academies and the Institute of Medicine. Dr. Bruce M. Alberts and Dr. William A. Wulf are chairman and vice chairman, respectively, of the National Research Council.

Acknowledgments

In accordance with NRC report review policies, this report has been reviewed by individuals chosen for their diverse perspectives and technical expertise. This independent review provided candid and critical comments that assisted the authors and the NRC in making the published report as sound as possible and ensured that the report meets institutional standards for objectivity, evidence, and responsiveness to the study charge. The content of the review comments and draft manuscript remain confidential to protect the integrity of the deliberative process. The Committee and staff wish to thank the following individuals for their participation in the review of the report: David Farmer (Institute of Ocean Sciences, Canadian Department of Fisheries and Oceans), Jonathan Gordon (Oxford University and International Fund for Animal Welfare), Robert Hofman (U.S. Marine Mammal Commission), Glenis Long (Purdue University), James Lynch (Woods Hole Oceanographic Institution), Donald Malins (Pacific Northwest Research Institute), Andrew Solow (Woods Hole Oceanographic Institution), and Bernd Würsig (Texas A&M University). While these people provided many constructive comments and suggestions, responsibility for the final content rests solely with the authoring committee and the NRC.

Contents

Marine Mammals and
Low-Frequency Sound

Executive Summary

Sound has become a major tool for studying the ocean. Although the ocean is relatively opaque to light, it is relatively transparent to sound. Sound having frequencies below 1,000 Hertz (Hz) is often defined as low-frequency sound. The speed of sound is proportional to the temperature of the water through which it passes. Therefore, sound speed can be used to infer the average temperature of the water volume through which sound waves have passed. The relationship between water temperature and the speed of sound is the basis for the Acoustic Thermometry of Ocean Climate (ATOC) experiment. The ATOC experiment is designed to monitor the travel time of sound between sources off the coasts of Hawaii and California and several receivers around the Pacific Ocean in order to detect trends in ocean temperature and for other research and monitoring purposes (ATOC Consortium, 1998). The ATOC transmissions are centered at a frequency of 75 Hz, with peak source levels of 180 decibels (dB) re 1 µPa @ 1m[1] at this frequency and 195 dB for its broadband source level. Based on well-tested models of signal loss over distance in deep water, the source level should decrease to 155 dB within 100 m from the source and to 135 dB at 1 km from the source.

Some whales, seals, and fish use low-frequency sound to communicate and to sense their environments. For example, baleen whales and some toothed whales are known to use and respond to low-frequency sound emitted by other

---

[1] Decibels are used to describe the ratio between two quantities, in this case, the ratio of the sound pressure level (SPL) of the source to the SPL of 1 microPascal (µPa) at one meter from the source. "re" = "with reference to." The report will henceforth omit the "re 1 µPa @ 1m" notation when referring to decibel levels measured in water. Measurements made in air are referenced to 20 µPa @ 1 m and will be identified in the text.

individuals of their species. Sharks are not known to produce low-frequency sound but are attracted to pulsed low-frequency sounds. Therefore, it is possible that human-generated low-frequency sound could interfere with the natural behavior of whales, sharks, and some other marine animals.

In 1994 the Defense Advanced Research Projects Agency (DARPA) requested that the National Research Council (NRC) convene a committee to evaluate the results of ATOC's Marine Mammal Research Program (MMRP) (see Appendix A for Committee biographies). The MMRP was designed to monitor the effects of ATOC transmissions on marine mammals. Although DARPA was the original sponsor, the Office of Naval Research (ONR) is now funding the MMRP and cosponsored this study. The NRC was asked to

1. conduct an updated review of *Low-Frequency Sound and Marine Mammals: Current Knowledge and Research Needs* (NRC, 1994), based on data obtained from the MMRP and results of any other relevant research, including ONR's research program in low-frequency sound and marine mammals;

2. compare new data with the research needs specified in the 1994 NRC report, focusing on the strengths and weaknesses of the data for answering the important outstanding questions about marine mammal responses to low-frequency sound; and

3. identify areas where gaps in our knowledge continue to exist.

The Committee reviewed numerous written documents and was briefed on the MMRP's progress at the program's midpoint (in 1996) and about 6 months after the completion of its field observations (in 1999). The NRC was asked to prepare an interim report to provide midproject guidance to the MMRP as well as this final report. Some of the recommendations in the interim report (NRC, 1996) were implemented by 1999, but for a variety of reasons others were not.

For its update of research priorities related to marine mammals and low-frequency sound, the Committee augmented the MMRP results with results from the scientific literature, ONR's program on marine mammals, and observations of the reactions of marine mammals to tests of the Navy's low-frequency active (LFA) sonar. This report does not examine the effects of all human-generated sound (only low-frequency sound is considered), nor does it include all marine mammals (only whales and seals are included). This report updates all aspects of NRC (1994), including the issue of acoustic harassment and its regulatory definition in light of the 1994 reauthorization of the Marine Mammal Protection Act (MMPA). The publication of the report is particularly timely because the MMPA expired on October 1, 1999 and is in the process of being reauthorized. The Committee focused exclusively on whales and seals, because (1) they are found near the ATOC sources, (2) the effects of low-frequency sound on whales and seals have been studied to a greater extent than effects on other marine mammals (in part, because they live near ATOC sources), and (3) it is thought that low-frequency sound is less likely to have a significant impact on other marine mammals, including sea and marine otters, manatees and dugongs, and polar bears.

FINDINGS AND RECOMMENDATIONS

Some of the MMRP observations, such as movements of humpback whales in near-coastal areas off Kauai and the abundance of some whale species near the Pioneer Seamount source off California, showed no statistically significant effects of ATOC transmissions. For these observations, the Committee could not distinguish among true lack of effect and insufficient observations, small sample sizes, and incorrect statistical treatment of data. A somewhat clearer lack of significant effects of the ATOC transmissions was demonstrated in observations of elephant seals' diving behavior near the Pioneer Seamount source. Some statistically significant differences between control and exposure conditions were found for other species, including (1) an increase in average distance of humpback whales from the California source and (2) increased dive duration for humpback whales off Hawaii. The MMRP found no obvious catastrophic short-term effects as a result of transmissions from either source, such as mass strandings or mass desertions of source areas.

Statements about whether ATOC should be allowed to continue, based on MMRP and other results, are clearly outside the Committee's statement of task. However, the Committee does offer suggestions about how future large-scale acoustic tomography experiments could be designed to accomplish appropriate monitoring for scientific purposes and mitigation measures to decrease the possibility of harm to marine mammals.

Progress has been made since 1994 in answering several of the research questions described in the 1994 NRC report. Research funded by ONR and other agencies and the results of the MMRP and LFA tests have contributed new knowledge regarding the effects of low-frequency sound on marine mammals. Research and observations published since 1994 have extended our knowledge of the hearing abilities of marine mammals at lower frequencies, at depth, in the presence of human-generated noise, and among different individuals of the same species. More observations of baleen whale vocalizations and responses to sound have been collected and a greater appreciation has been gained about how the respective locations of a baleen whale and a sound source can affect vocalizations and other behavior. Extensive testing with conventional and new methods, such as computational modeling of ear anatomy, auditory evoked potential techniques, and stimulus-response experiments with trained animals have provided new insights into normal hearing and the levels of sound required to produce shifts in the hearing abilities of individual animals.

Most of the research directions recommended by the 1994 report are still relevant. This continued need to answer the questions raised therein is not due to lack of effort but is a result of the complexities of the questions and the difficulties of conducting studies on marine mammals because of the lack of adequate research support, equipment, techniques, and facilities.

The 1994 amendments to the MMPA (16 U.S.C. 1361 et seq.) changed the legal definitions of marine mammal "harassment" as applied to scientific use of sound in the ocean. If the MMPA is to be implemented responsibly, however, additional changes should be made to the act and to the regulations promulgated pursuant to the act by the Office of Protected Resources of the National Marine Fisheries Service (NMFS) in the National Oceanic and Atmospheric Administration (NOAA).

There is little disagreement that scientific use of sound in the ocean is a minor component of human-generated sound pollution. Industry (e.g., shipping and hydrocarbon exploration and production) are thought to be the largest sources. Yet, uses of sound by scientists and the Navy are the most stringently regulated. Unfortunately, few data are available to regulators regarding ambient noise levels in the ocean and the relative importance of different sources in contributing to the cumulative human-generated noise. Cooperative funding of research by government and industries responsible for the noise could result in more rapid advance of knowledge about the effects of sound on marine mammals and cooperative solutions to noise problems.

This report includes a number of recommendations to Congress, to NOAA in its regulatory role, and to research sponsors, as well as to the scientific community. The recommendations directed to Congress should be implemented in the upcoming reauthorization of the MMPA. The recommendations directed to NOAA in its regulatory role should be implemented as it promulgates new regulations based on the reauthorized MMPA. Finally, agencies that fund marine mammal and acoustic research should begin weighing recommendations about research, monitoring, and facilities against other budget priorities for the fiscal year 2002 budget cycle and beyond. Some of the recommendations to research sponsors should not require reprogramming or new money and could be implemented immediately.

Recommendations for Congress

As part of the upcoming reauthorization, Congress should consider changes to the MMPA that would allow studies of the ocean while protecting marine mammals. In particular, Congress should consider the following actions:

- define "type B harassment" of marine mammals in terms of significant disruption of behaviors critical to survival and reproduction.
- acknowledge the relative significance of different sources of sound in the ocean, insofar as this is known, and provide new means to bring all commercial sources of sound into the MMPA's legal and regulatory framework.

The committee believes that regulation of sound in the ocean is based on inadequate information and that more information needs to be collected. Congress should decide what kinds of regulations are appropriate and how much funding should be available for marine mammal research, given the existing inadequacy of knowledge.

Recommendations for NOAA

NOAA's responsibilities with respect to whales and seals are set forth in the MMPA, the Endangered Species Act, and other relevant legislation. NOAA's responsibility has been delegated to NMFS. Although NMFS conducts and supports some marine mammal research, it has conducted or supported very little research aimed at determining the potential effects of anthropogenic sound on the distributions, sizes, or productivity of marine mammal species or stocks. In September 1998, NMFS held a workshop to seek input from the scientific community regarding guidelines or regulations that might be promulgated to guide or govern authorization of the taking of marine mammals incidental to activities that use or produce sound in the ocean (no publication resulted from the meeting). The workshop participants noted a variety of uncertainties concerning the possible effects of anthropogenic sound on marine mammals. Pending resolution of the uncertainties, NMFS should focus on developing and evaluating the effectiveness of guidelines for preventing injuries and disruption of behavior that could affect survival or reproduction. NMFS should consult further with experts in oceanography, bioacoustics, underwater sound propagation, and animal behavior to (1) identify sound-producing activities that, because of their nature, location, intensity, or duration, are likely to have biologically significant effects on marine mammals and thus should be higher priority for enforcement of the "taking" authorization under the MMPA or the Endangered Species Act; and (2) for cases in which there is uncertainty or disagreement as to possible adverse effects of underwater sound on survival or productivity, describe (a) the research required to resolve the uncertainty, and/or (b) the monitoring that should be required as a condition of any incidental take authorization provided by NMFS. Further, NMFS should work cooperatively with ONR to develop technology and programs for monitoring ambient sound levels and noise pollution in critical marine mammal habitats and to develop and implement methods for obtaining data on the hearing capabilities of marine mammals, including data on auditory sensitivity, damage thresholds, and potential for behavioral disruptions of representatives of all types of marine mammals (see Box 5.1).

Recommendations for Research Sponsors

Developing an understanding of the effects of low-frequency sound on marine mammals will require a more sustained and integrated approach than has been the case in previous research. Much research in the past was conducted by single investigators responding to the need for specific information about the effects of a single sound source. Multi-investigator teams of biologists, acousticians, psychoacousticians, engineers, and statisticians should be funded to conduct a set of systematic studies of marine mammal species that represent different potential hearing abilities, based on the need to know how sound of different types affects characteristic species. The committee also identifies the need for research to determine:

• how marine mammals utilize natural sound for communication and for maintaining their normal behavioral repertoires;
• the responses of free-ranging marine mammals to human-generated acoustic stimuli, including repeated exposure of the same individuals to the same stimulus;
• the response of deep-diving marine mammals to low-frequency sounds whose characteristics duplicate or approximate those produced by acoustic oceanographers and other sources of human-generated sound, such as low-frequency military sonars and sounds used for seismic exploration;
• basic hearing capabilities of various species of marine mammals;
• hearing capabilities of larger marine mammals that are not amenable to laboratory study;
• audiometric data on multiple animals of different sexes and ages in order to understand variance in hearing capabilities within a given species;
• sound pressure levels that produce temporary and permanent hearing loss in marine mammals;
• condition of a representative sample of important cochlear structures in different species of wild marine mammals using post-mortem examinations;
• morphology and sound conduction paths of the auditory system in various marine mammals;
• temporal-resolving power for various marine mammals;
• whether low-frequency sounds affect the behavior and physiology of organisms that serve as part of the food chain for marine mammals; and
• whether low-frequency sounds affect the nonauditory physiology or structures of marine mammals.

Such research should be sponsored by the agencies that fund basic and applied biological research and that fund ocean research using sound, including ONR, NOAA, the National Science Foundation (NSF), the Minerals Management Service, the Biological Resources Division of the U.S. Geological Survey, and the National Institutes of Health (NIH). Mission-oriented agencies should ensure that the research they sponsor will not only contribute to their immediate missions but also answer basic questions about marine mammal bioacoustics. Agencies that fund more fundamental science, such as NSF and NIH, should consider funding marine mammal research when it has implications for understanding basic biology or health-related issues. Most importantly, all of these projects should receive strict peer review and be evaluated on the quality of the science proposed.

Other generators of sound in the ocean, such as shipping and hydrocarbon exploration and production companies, also should participate in funding research on the effects of sound on marine mammals. Given our ignorance about safe exposure levels of sound, great benefit could accrue if ocean noise generators, government agencies, and environmental groups formed a consortium to fund the kinds of research recommended in this report. Opportunities may also exist for cooperation between U.S. scientists and agencies and their counterparts in other nations. Cooperation with Canada and Mexico could be particularly productive because several species cross the exclusive economic zones of the three nations. For example, another NRC (1999) report described research on marine mammals that could benefit from binational research by the United States and Mexico. Europe is also a likely source of partners for cooperative research and management, given the shared marine mammal stocks and the existing cooperation in the North Atlantic Treaty Organization, which shares both active and passive sonar sources with the United States. A variety of organizations, including the Ocean Drilling Program, provide models for the possible structure and functioning of a multinational consortium for research on the effects of sound on marine mammals.

Research on captive marine mammals is expensive because of the need for extended training and maintenance of animals and the added requirement of highly specialized care (e.g., aquatic veterinarians). Funds to support marine mammals must be provided for the long term because once an animal is in captivity it generally must be maintained there for its lifetime. Facilities to conduct research with marine mammals are difficult to set up, and most existing commercial facilities are not able to provide access to animals for research. However, without such facilities, many basic science studies on marine mammal bioacoustics (and other aspects of marine mammal biology) such as those described in this report cannot be conducted, and it will be difficult to develop regulations that protect marine mammals appropriately. The lack of a specialized marine mammal research facility available to U.S. scientists has hindered the progress of research on marine mammal hearing. If the studies described in this report are of sufficient priority to reduce uncertainties in the regulation of human-generated sound in the ocean, federal agencies should consider establishing a national facility for the study of marine mammal hearing and behavior. If established, the proposed facility should be made available to the entire scientific community, and the allocation of animal experimental and observation time should be based on the scientific merit of proposals as determined by peer-reviewed evaluation of research. Funding for research at this facility should be coordinated with the availability of animals to ensure that once an investigator receives funding he or she will have access to appropriate animals. The committee believes that such a facility could be established at relatively little incremental cost by enhancement of an existing facility.

Our understanding of how marine mammals react to natural and human-made sound is rudimentary. The actions recommended in this report could result in significant advances in knowledge and better regulation of human activities that might be harmful to marine mammals.

1   Introduction

Sound is an important tool used by ocean scientists to study the topography of the seafloor and its substructure; the direction and speed of ocean currents; and the size, shape, and number of organisms in the ocean. Four fundamental properties of sound transmission are important to understand as background for this report:

1. The transmission distance of sound in seawater is determined by a combination of geometric spreading loss and an absorptive loss proportional to the sound frequency (see Box 1.1). Thus, attenuation of sound increases as its frequency increases.

2. The speed of sound is proportional to the temperature of the seawater through which it passes.

3. The sound intensity decreases with distance from the sound source. Generally, the decrease in sound intensity ranges between $1/r$ (r = distance from the source) and $1/r^2$ (spherical spreading), depending on characteristics of the sound source location and transmission paths, although sound intensity can decrease even more under certain conditions. Thus, a sound level may be as much as 60 dB lower than that of the source level at 1 km from the source (see Figure 1.1). Because of the wave properties of sound and propagation conditions, waves from different sources or refracted and reflected waves from a single source can converge and either add to or cancel each other, so that simple geometric models of spreading do not always predict actual sound fields in the ocean. This is especially true in shallow water.

BOX 1.1 Transmission Loss

Transmission loss is defined as 10 times the logarithm to the base 10 of the ratio of the intensity at 1 m to the intensity at the range of interest: TL (transmission loss) = $10 \log_{10}(I_{1\,m}/I_{range\,point})$. If the loss is geometric, it can involve either spherical spreading $\rightarrow TL = 20 \log_{10}R$, where R is the distance from the source in meters or cylindrical spreading $\rightarrow TL = 10 \log_{10}R$.

In general, a "rule of thumb" is used, which is spherical spreading and attenuation, so $TL = 20 \log_{10}R + \alpha R$, where α is proportional to f^2 (f = frequency). In general, α is quite complex, being made up of viscous losses, heat conduction losses, relaxation losses, and other losses, but it is a good approximation to say that it is directly proportional to f^2. At ATOC transmission frequencies, α is so small that it only contributes to transmission loss significantly at long distances.

---

4. The strength of sound is measured on a logarithmic scale, $10 \log_{10} I/I_{ref}$ (I = intensity[1]); therefore, 180 dB is 10 times less intense than 190 dB, and 170 dB is 100 times less intense than 190 dB.

Because of property 1, only low-frequency sounds are useful for studying large-scale processes over long distances, such as the structure of the ocean over scales of hundreds to thousands of kilometers. For example, sound has been used to study circulation patterns in the ocean using tomographic techniques analogous to the CAT [computerized axial tomography] scan technology used in medicine (Munk and Wunsch, 1979; Munk et al., 1995). Likewise, sound is used in geophysical studies to characterize the subsurface structure of the seafloor. The decrease in sound transmission distance with increasing frequency also has implications for marine mammal communication because only low-frequency vocalizations can travel long distances. Because of property 2, sound speed can be used to infer the average temperature of the water volume through which the sound waves have passed. Scientists are using the relationship of the speed of sound and water temperature to infer whether global warming is occurring. The Acoustic Thermometry of Ocean Climate (ATOC) experiment is monitoring the

---

[1]Intensity is considered the fundamental quantity of sound, but it is seldom measured. Instead, pressure is normally measured. The two are related by $I = p^2/\rho_0 c_0$, where p is the time-averaged pressure, ρ_0 is the density of the medium, and c_0 is the sound speed in the medium. An acoustic wave whose pressure is 1 µPa has an intensity of $0.64 * 10^{-22}$ watts/cm². For transient signals, it is more meaningful to refer to the energy flux density (E) of the acoustic wave. The energy flux density is the time integral of the instantaneous intensity.

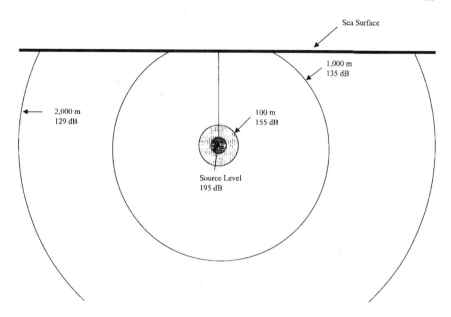

FIGURE 1.1   Calculated received levels at various distances from the ATOC source, based on spherical spreading, assuming no reflections from the sea surface or bottom. Received levels are affected by conditions in the ocean, where the source is deployed (on the bottom or in the water column) and thus reflection from the sea surface and the seafloor, and directionality of the source. Spherical spreading is a proper assumption at these distances until sound waves reach a boundary (see Urick, 1983).

travel time of sound between sources off the coasts of Hawaii and California to receivers around the Pacific Ocean (see Figure 1.2) for a variety of purposes (see section below on "The ATOC Concept").

Ambient noise levels vary both from place to place in the ocean and over time at each location. The relative frequency bands also vary, due in part to the nonrandom distribution of vocal animals and human-generated noise. Measurements and predictions of ambient noise in the ocean were made by Knudsen et al. (1948) and Ross (1976). Natural ambient noise levels increase as frequency decreases and are related to the sea state. Ross (1976) reported that the ambient noise in areas of heavy shipping activity could range between 85-95 dB (1 Hz bandwidth), peaking at a frequency of about 100 Hz. High sea states can produce similar levels of ambient noise.

Some whales, seals, and fish use low-frequency sound to communicate and to sense their environments (Tyack, 1998). For example, baleen whales and

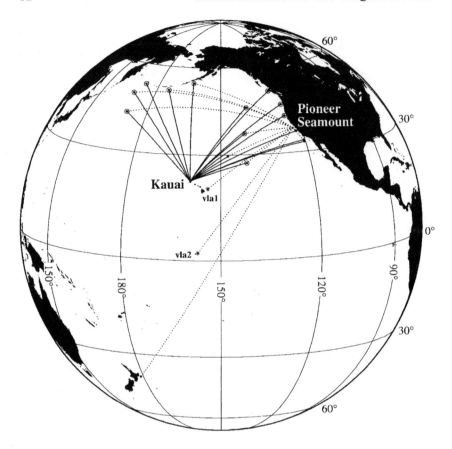

FIGURE 1.2 ATOC sources at Pioneer Seamount and Kauai, showing transmission paths to receivers in different parts of the Pacific Ocean.

some toothed whales are known to use and respond to low-frequency sound emitted by other individuals of their species (McDonald et al., 1995; Edds-Walton, 1997; Ljungblad et al., 1997; Stafford et al., 1998). Sharks and some other fish species are able to sense and react to low-frequency sound (Myrberg et al., 1976; Myrberg, 1990). Therefore, it is possible that human-generated low-frequency sound can interfere with the normal behavior of some marine animals and there is some evidence that this occurs (Myrberg, 1978, 1980, 1990; Richardson et al., 1995). Serious misunderstandings of the potential effects of sound of various intensities on marine mammals have occurred because the levels of sound intensity in water and in air have not been consistently (or in some cases, correctly) referenced to the International System of Units (SI) standards[2] that have been established and in-water sound levels have been misunderstood to be comparable directly to in-air levels, with which most people are more familiar. Air-water comparisons are inherently misleading.

THE ATOC CONCEPT

Individuals studying the ocean or using it for scientific, commercial, or military purposes use underwater sound as a major tool to monitor and explore the ocean's contents and boundaries. Sound also enters the ocean as a result of natural environmental processes, biological activity, and human activities unrelated to study of the ocean, such as the propulsion noise of ships (Table 1.1). Although all kinds of sounds are used, many applications have used sound in the 1- to 100-Hz frequency range because absorption of these sound frequencies by seawater is minimal, variability caused by the environment is somewhat reduced, and long-range propagation is possible, making underwater or subbottom remote sensing feasible.

An example of the scientific use of low-frequency sound in the ocean was the Heard Island Feasibility Test (HIFT), in which sound was transmitted from one array of sources with a sound level of 221 dB (rms)[3] and a frequency of 57 Hz (rms) through the ocean to a number of receivers over distances of up to 16,000 km (Baggeroer and Munk, 1992). A major goal of HIFT was to serve as a prototype for regular observations of the speed of sound in the ocean as a direct means of measuring the rate of ocean warming due to global climate change. The regular observations were proposed as the long-term ATOC experiment.

The issue of global warming is of major significance to scientists, policymakers, and citizens worldwide, yet it has been difficult to determine the extent of atmospheric and oceanic warming based on observations of global air and sea surface temperatures.[4] The advantages of long-distance sound transmissions in the ocean are that (1) low-frequency sound waves pass through and thus sample a wide range of ocean depths between the source and the receiver(s), (2) the summed effects of random variability along the transmission path due to eddies and variations in ocean currents are minimized, and (3) longer-distance

---

[2]ANSI S1.8-8-1989 (ASA 84-1989), Revision of S1.8-1969 (R 1974), Reaffirmed by ANSI on July 29, 1997. In the International System of Units (SI), acoustic pressure is expressed in watts per square meter, but the dB notation is used more commonly at the present time.

[3]The amplitude of pulsed sounds is typically expressed as "peak-to-peak" (one cycle of the sine wave) or "zero-to-peak" (one-half cycle). Continuous sounds may be expressed as "root-mean-square" (rms), which is the square root of the time average of the square of a quantity; for a periodic quantity the average is taken over one complete cycle (Lapedes, 1974).

[4]The NRC recently issued a report on the measurement of atmospheric global warming (NRC, 2000).

TABLE 1.1   List of Some Anthropogenic Sounds, Including Sources, Frequencies, and Levels

Source	Frequency at Highest Level 1/3-Octave Band (Hz)	Source Level at Highest Level 1/3-Octave Band (dB re 1 μPa @ 1 m)
5-m Zodiac inflatable boat[a]	6,300	152
Bell 212 helicopter[b]	16	159
Large tanker	100 + 125	177
Icebreaker	100	183
ATOC	75	195[c]
Air gun array (32 guns)	50	210[d]
HIFT	50 + 63	221[e]
Military search sonar	2,000-5,000	230+

SOURCE: Richardson et al. (1995, Table 6.9).

 [a]Speed and horsepower of engines were not given in Richardson et al. (1995).

 [b]Aircraft flyover source levels were computed by Malme et al. (1989) for a standard altitude of 1,000 ft (305 m). For consistency with other sound sources, these values were changed to a reference range of 1 m by adding 50 dB.

 [c]Numbers provided by ATOC investigators from actual transmissions, rather than from Richardson et al. (1995).

 [d]Anecdotal evidence suggests that airgun arrays may reach source levels of 240 dB.

 [e]Numbers provided by Heard Island Feasibility Test (HIFT) investigators.

transmissions may allow the detection of smaller changes in temperature. The designers of ATOC hope to conduct ATOC transmissions from Kauai for at least 5 additional years to make a quantitative assessment of the role that acoustic thermometry can play in an integrated ocean-observing system for ocean weather and climate in the North Pacific Ocean (P. Worcester, Scripps Institution of Oceanography, personal communication, 1999). Results from the initial ATOC transmissions indicate that this technique shows promise for at least one of its planned applications, namely, to use ocean temperature measurements from ATOC to constrain climate models (ATOC Consortium, 1998). ATOC Consortium investigators compared sea-level estimates derived from historic averages, ATOC-based tomography, recent direct measurements, results from a general circulation model (GCM), and data from a satellite-based altimeter. Combinations of the GCM, altimeter, and ATOC data show that the GCM alone underestimates the magnitude of the seasonal sea surface heat flux cycle. However, despite the usefulness of acoustic tomography programs like ATOC and other research uses of low-frequency sound in the ocean, concern exists that adding more sound to the ocean could harm marine mammals, sea turtles, and other organisms, as the following section will describe.

It is necessary to sample the ocean's temperature frequently enough to be able to distinguish any trend of temperature increase amidst the "noise" of random variations created by temperature, tides, and mesoscale ocean structure.[5] Information gained from ATOC-like techniques cannot be replaced by measurements from satellites because satellites can only sense features of the surface layer of the ocean. The temperatures from the full depth of the ocean can be measured *in situ*, but such measurements are limited in number and frequency because of the cost and limited number of oceanographic ships, moorings, and drifters available, compared to the great volume of the ocean. The planned Array for Real-time Geostrophic Oceanography (ARGO) drifter program will provide a new and innovative means of measuring ocean interior temperatures over large scales (albeit not in an integrated, synoptic manner), and may provide a complement to, and possibly a replacement for, ATOC-type acoustic measurements of water temperature.

LOW-FREQUENCY SOUND AND MARINE VERTEBRATES

It is well known that some marine organisms produce low-frequency sounds and/or can hear such sounds. For example, there is evidence that baleen whales (such as **finback** [*Balaenoptera physalus*], **blue** [*Balaenoptera musculus*], and **humpback** whales [*Megaptera novaeangliae*]) communicate using low-frequency sound (e.g., reviewed in Edds-Walton, 1997). Table 1.2 shows the frequency range and dominant frequencies of the vocalizations of a sample of baleen whales, toothed whales, and seals. The geographic extent of the use of low-frequency sounds by baleen whales is being monitored on an experimental basis in the Atlantic and Pacific oceans using a novel source of data—the Integrated Undersea Surveillance System (IUSS) formerly known as the SOund SUrveillance System (SOSUS)—which was originally designed for tracking submarines. The IUSS has allowed the tracking of individual whales in at least a few cases by triangulating the positions of vocalizations over time (Clark, 1995; Stafford et al., 1998; Watkins et al., 2000). Such data are important in determining the migration behavior of individual whales and in assessing whether human influences change these pelagic migrations. Richardson et al. (1995) present a comparison of the audiograms of some species of marine mammals (see Figure 1.3). Addi-

---

[5]As Peter Worcester, ATOC principal investigator, explained to the Committee in writing in 1999 "The required duty cycle is actually set by the need to avoid aliasing of rapidly changing oceanographic phenomena. If high-frequency phenomena are sampled at too low a rate, they will erroneously appear in subsequent analyses as low-frequency variability. In general one needs to sample at more than twice the highest frequency containing significant energy to avoid aliasing. In the case of the ocean, mesoscale variability has timescales from a week to a few months, and so needs to be sampled every few days. The tides are of course even higher frequency, but because their frequencies are well known, they can be sampled adequately using a frequency of approximately 1 day out of every few days. This combination of ocean phenomena led to a 2 percent duty cycle being used, consisting of 1 day with six 20-minute transmissions at 4-hour intervals to adequately sample tidal variability, occurring every fourth day to adequately sample ocean mesoscale variability."

TABLE 1.2   Frequencies Used in Communication and Echolocation by
Selected Marine Mammals

Species	Frequency Range (Hz)[a]	Dominant Frequencies (Hz)[b]
Selected Baleen Whales		
Gray Whale		
adults	20-2,000	20-1,200
calf clicks	100-20,000	3,400-4,000
Humpback Whale	30-8,000	120-4,000
Finback Whale	14-750	20-40[c]
Minke Whale	40-2,000	60-140[d]
Southern Right Whale[e]	30-2,200	50-500
Bowhead Whale	20-3,500	100-400
Blue Whale		
Atlantic[f]	—	10-20[h]
Pacific[g]	10-390	16-24
Selected Toothed Whales		
Sperm Whale (clicks)	100-30,000	2,000-16,000
White Whale[i]		
whistles	260-20,000	2,000-5,900
clicks	40,000-120,000	—
Killer Whale		
whistles	1,500-18,000	6,000-12,000
clicks	1,200-25,000	—
Bottlenose Dolphin		
whistle	800-24,000	3,500-14,500
clicks[j]	1,000-150,000	30,000-130,000
Selected Seals		
California Sea Lion (in air)	<1,000-<8,000	500-4,000
Harbor Seal (in air)	<100-150,000+	<100-40,000
Gray Seal	100-40,000	100-10,000

SOURCE: Modified from Richardson et al. (1995).

[a]The frequency range listed is the lowest to highest frequencies listed by Richardson et al. (1995) and more recent authors. Gaps in the ranges are not shown.

[b]Dominant frequencies are essentially the bandwidth of sound that has the greatest energy. They do not include all the frequencies produced, since there may be many weak harmonics.

[c]Edds (1988).

[d]Edds-Walton (2000).

[e]Although few recordings exist, the northern right whale repertoire is likely to be similar.

[f]Published data are too limited to give the frequency range for this population.

[g]Stafford and Fox (1996).

[h]Edds (1982).

[i]W. Perrin and D.W. Rice, both NMFS experts in taxonomy, verified that individuals of the species *Delphinapterus leucas* can be called white whales, belugas, or belukhas. White whale is used throughout this report.

[j]Ridgway and Au (1999).

tional information about the effects of low-frequency sound on marine mammals is contained in Chapters 2 and 3.

Low-frequency sounds are used by other marine vertebrates, including sharks and bony fish (Myrberg, 1972, 1978, 1980, 1990). Sharks are attracted to sources emitting such sounds as possible food indicators (e.g., Myrberg, 1978), and many species of fish use low-frequency sounds for communication (e.g., Demski et al., 1973).

ORIGIN OF STUDY

As a result of issues raised by HIFT, the Office of Naval Research (ONR) requested in 1992 that the National Research Council examine the state of knowledge of the effects of low-frequency sounds on marine mammals and assess the trade-offs between the benefits of underwater sound as a research tool and the possible harmful effects on marine mammal populations of introducing additional low-frequency sound into the ocean. In 1994 the NRC issued a report, *Low Frequency Sound and Marine Mammals: Current Knowledge and Research Needs,* which concluded that (1) very little is known about the effects of low-frequency sound on marine mammals and (2) it is difficult to establish regulatory policy in the absence of data regarding such effects (see Appendix B for the executive summary from that report). The report included a series of recommendations about the kinds of research needed to fill the gaps in our knowledge.

Subsequent to HIFT, the ATOC program was proposed with a mission to make regular measurements of the travel times of low-frequency sound throughout the Pacific Ocean (Figure 1.2). As a result of concerns about the effects of low-frequency sound added to the ocean by ATOC, the ATOC program conducted the first several years of transmissions under a permit to test the effects of ATOC sound sources on marine mammals through a Marine Mammal Research Program (MMRP). The Defense Advanced Research Projects Agency requested that the NRC update the information contained in its 1994 report based on the MMRP and other results.[6] In addition, the NRC was asked to ascertain how data acquired since 1994 fulfill the research needs described in the 1994 report. An interim NRC report published in 1996 provided guidance to the MMRP in the midst of its observational studies. The director of the ONR program (Robert Gisiner) and the principal investigators of the MMRP (Christopher Clark and Daniel Costa) briefed the NRC's Committee to Review Results of ATOC's Marine Mammal Research Program in 1996 and 1999 and participated in subsequent open discussions.

The Committee summarizes and comments on the results of the MMRP in Chapter 2. Chapter 3 is devoted to updating the research priorities first identified

---

[6]The MMRP formed its own advisory board to provide independent advice to MMRP investigators regarding MMRP needs, plans, schedules, and research results.

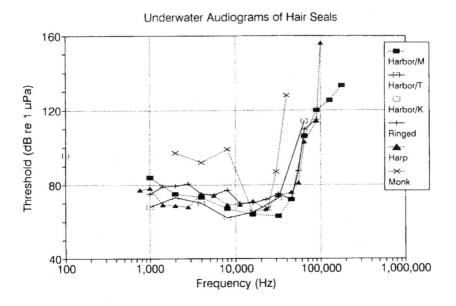

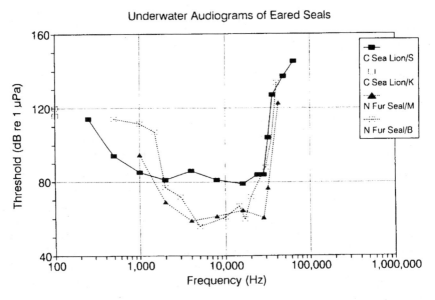

FIGURE 1.3 Audiograms of representative seal and toothed whale species. Source: Richardson et al. (1995); used with permission from Academic Press. References for these data are given in the source document. In most cases, the data represent measurements on one or two individuals of the species.

NOTE: Complete audiograms should be U-shaped. If not, hearing should be tested at higher or lower frequencies, as necessary. For example, the audiograms of the true seals (shown as "hair seals" in the figure) appear to be truncated at lower frequencies.

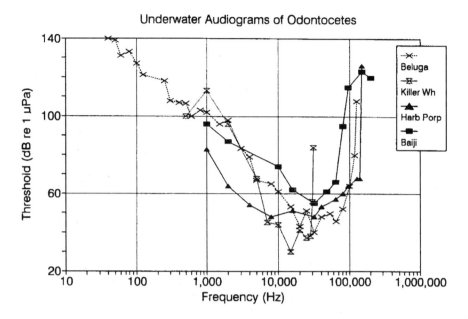

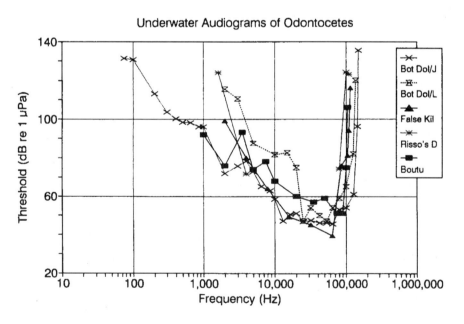

in the 1994 NRC report, based on data obtained from research conducted by the MMRP, as well as the results of other relevant research such as that sponsored by the ONR program on marine mammals and ocean acoustics. Based on this comparison of recent research achievements and research needs listed in the 1994 report, the present report identifies areas in which gaps in our knowledge continue to exist. Chapter 4 discusses regulatory issues, such as how permits for acoustic and marine mammal research are issued. Chapter 5 draws together the Committee's findings and provides recommendations based on these findings.

2 Evaluation of the Marine Mammal Research Program

DESCRIPTION OF MMRP RESULTS

The sound source used by the Acoustic Thermometry of Ocean Climate (ATOC) experiment is acoustically different from the source used in the previous Heard Island Feasibility Test (HIFT), in part due to concerns about potential effects on marine mammals and in part to the shorter distance between the ATOC sources and receivers. In fact, information obtained from HIFT indicated that a less intense sound source (HIFT used a level of 221 dB) could be used for ocean basin-scale studies such as the ATOC experiment (Baggeroer and Munk, 1992). The ATOC source level was thus reduced to 195 dB. This difference corresponds to a 400-fold decrease in sound intensity. The 75-Hz ATOC signal was transmitted from sources located at 850- and 980-m depths off the coasts of Hawaii and California, respectively, for 20 minutes every 4 hours every fourth day, under the standard protocol of a 2 percent duty cycle (see footnote on page 15). This standard protocol was varied somewhat in both the California and the Hawaii transmissions, depending on the needs of MMRP investigators, although the transmissions were not optimized for studies of marine mammals. The California source transmitted for experimental periods of 2 to 4 days, separated by at least 4 days with no transmissions. The Kauai source used a similar protocol during the first season, followed by the standard protocol of 1 day of transmissions every 4 days. One exception was that the duty cycle of the Kauai source was increased to 8 percent in the summer of 1999, after the humpback whale season, in accordance with the environmental impact statement (EIS) for the Kauai source (ARPA and NOAA, 1995).

For all ATOC transmissions, the source level in dB increased linearly from 165 to 195 dB and the power increased logarithmically over a period of 5 minutes preceding the 20-minute, full-intensity transmission. This ramp-up period was designed on the assumption that it would allow marine mammals the opportunity to move away and avoid exposure to the sound if it annoyed them.[1] As a result of the decrease in source level and the use of a sound ramp-up, the potential for acoustic impact on marine mammals presumably has been reduced in ATOC compared with HIFT. The sound level at the 850- to 980-m deep sources should diminish to approximately 130 dB at the sea surface, plus or minus a small fraction of this value due to the Lloyd mirror effect.[2] Thus, marine mammals that spend most of their time in surface waters potentially are exposed to much lower received sound levels[3] than the source level, although deep-diving species could be exposed to higher levels when diving near the source. Geometric spreading also diminishes sound levels in all directions from a source, so that the received level is expected to be about 135 dB at a radial distance of 1 km from the source and 129 dB at 2 km (see Figure 1.1).

The Committee assessed all available information and concluded that the Marine Mammal Research Program (MMRP) was not able to demonstrate a lack of significant effects of ATOC transmissions on marine mammals. The MMRP did not provide unequivocal evidence about the effects of ATOC transmissions because (1) the MMRP data were not fully analyzed as of April 1999 and (2) several of the observational programs were not designed in accordance with the suggestions made in another NRC report (NRC, 1996) that may have helped reduce the ambiguity of the results. It would have been impossible for the MMRP program to conclusively demonstrate a lack of subtle or long-term effects within the short period of the program and the program did produce some useful results that advance our understanding of the effects of sound on marine mammals. However, it is important for those designing future studies to recognize that simply not detecting reactions is not by itself sufficient evidence that there is no significant impact.

---

[1]Costa and Williams (1999) estimated that sustained swimming speeds of many marine mammals are about 2 m/sec. Thus, the time needed for a marine mammal to swim from near the source to a distance at which the received level would be 120 db (5.6 km) would be 47 min. The time needed to reach the 130-dB received level distance (1.8 km) would be 15 min. Thus, the characteristics of the ATOC signal ramp-up period could expose marine mammals to levels of sound of 130 to 165 dB for periods of as much as 15 min.

[2]The Lloyd mirror effect creates a diminished or augmented pressure of sound from an underwater source either located near the water-air boundary or when received near that boundary. It is caused by the interference between direct and surface-reflected waves and thus creates alternating sound nulls and peaks around this level.

[3]The "received level" is the sound pressure level measured at the animal, that is, the level to which it is actually exposed. The received level is lower than the source level, depending on the distance between the source and the animal, the sound frequency, and environmental factors.

The MMRP was awarded $3 million to conduct its work over 5 years. The Committee did not examine how this funding was allocated to different activities, or whether the funding was adequate to meet the goals, staffing levels, or any other management matters, so it is impossible to determine whether the program was hampered by inadequate funding for the necessary tasks, poor planning and execution of observations, constraints placed on the program by the ATOC experimental design or regulatory requirements, the difficulty of working with large whales, or other factors. Although the MMRP observations did show some indications that the ATOC signal did not have a short-term effect on nearby populations of marine mammals and there were no obvious mass mortalities of marine mammals or abandonment of the ATOC source areas by marine mammal species under observation, there was little detailed observational evidence of the effect of the ATOC signal on individual whales. The MMRP results and the committee's evaluation of the significance of the results are given in Table 2.1.

The Committee makes a number of recommendations in Chapter 5, based on the MMRP experiences, about the need for peer-reviewed research, multidisciplinary research teams, proactive research programs not linked to specific acoustic experiments, and the need to devote sufficient financial and human resources to ensure timely data analysis and publishing of results. Because the MMRP did not provide unambiguous results about the effects (or lack thereof) of the ATOC transmissions, the Committee cannot state unequivocally whether or not ATOC transmissions should continue. (The Committee was not asked to make such an assessment, but the question arose in the natural course of the Committee's discussions.) Instead, in the event of ATOC continuation or other large-scale acoustic tomography experiments, the Committee offers some criteria that should be considered and some mitigation measures that may reduce concerns about such experiments.

California ATOC Source

The goals of the California portion of the MMRP included (1) sampling the distribution and abundance of marine mammals in the vicinity of the source, (2) testing for differences in those distributions when the source was on and off, and (3) measuring diving responses of a marine mammal (the **northern elephant seal**, *Mirounga angustirostris*) as it passed the source while returning to its rookery. Because of the distance of the source from land, shore-based observations were precluded, so the distributions of marine mammals were sampled using aerial surveys. Observations were also conducted from a small boat for photographic identification of humpback whales and blue whales and for enumeration of other marine mammals. Aerial surveys provided a more or less synoptic picture of whale distribution throughout the survey area in a single day. These observations were designed to test whether the distribution of whales around the sound source would differ when the source had been transmitting for 1 to 3 days,

TABLE 2.1   MMRP Experiments/Observations, Results, and Significance

Experiment/Observations	Results
California Results	
Naturally migrating elephant seals (Costa et al., 1999)	• Although the power of the test was low ($\alpha = 0.05$; power $= 0.178$), analysis of the five male seals returning when the source was on (2-9 km) versus the 11 control male seals (2-66 km) shows no significant difference in closest distance to the source ($t = -1.524$, $df = 14$, $p = 0.15$).
Translocated elephant seals (Costa et al., 1999)	• Data analysis is incomplete, but male seals exposed to received levels ranging from 118 to 137 dB in the 60- to 90-Hz band did not seem to change their dive patterns.
Distributions of marine mammals by aerial observations (Calambokidis, 1999)	• A high diversity of species (including six endangered marine mammal species) and numbers of individuals was observed both when the ATOC source was on and off. • Humpback whales within a 40 * 40 km inner box centered around the ATOC source were sighted on average 2.4 km farther from the sound source when it was on versus off. No significant difference was apparent in sightings of humpbacks in an 80 * 80 km outer box excluding the inner box. Sperm whales showed a similar response, but data are complicated by dependence of clustered sightings of subgroups. Risso's dolphins were also found farther from the source in the experimental condition within 24 hours of a transmission. • Behavior of marine mammals was generally similar comparing exposure to control conditions. No significant differences were observed in the orientation of humpback and sperm whales, but limited sightings of blue whales suggest they oriented more toward the sound source during the transmissions, a trend nearing the $p = 0.05$ significance level.

ATOC MMRP Conclusion	NRC Committee Conclusion

- Although the one migrating seal for which there was a dive record did not appear to react to the initiation of the sound, there does appear to be a deviation in the dive pattern that resulted from the cessation of the ATOC transmission.

- No evidence was obtained to indicate that the ATOC source disrupts the geographic locations of migrating elephant seals, but the power of the test is limited by small sample size and lack of data on female seals. More data are needed on influence of ATOC on dive patterns of migrating seals.

- The first significant finding was that the highest level of exposure to ATOC for seals intentionally placed near the source was 137 dB in the 60- to 90-Hz band.
- The second significant finding was the lack of a dramatic response to operation of the ATOC sound source in any seal, even though the seals could almost certainly hear the source.

- These are the only data on possible effects of ATOC on diving behavior. Statistical procedures did not correct for multiple tests; test results need to be reanalyzed before final conclusions can be drawn.

- No apparent differences in number of sightings comparing control and exposure periods, but there were statistically significant differences in the distribution of some species.
- While humpback whales did not vacate the area during ATOC transmissions, some whales shifted from using areas near the sound source (<14 km from the source) to areas slightly farther away (14-28 km from the source).

- Given the high diversity and number of animals, and the possibility that the Pioneer Seamount is a critical habitat area for marine mammals, it is probably not a good area for the ATOC source to be located for decades of operation. ATOC has discontinued transmissions from this source.
- Not enough data are available to determine whether most species showed avoidance or attraction. Significant differences in distance of humpback whales from the source indicate an avoidance response, but the scale of this response is small enough that this is unlikely to impact availability of habitat for the population.
- Aerial survey data suggest possible vertical avoidance response for some species. For example, there is some evidence for increased sightings of sperm whales during exposure, indicating that exposed whales might spend more time at the surface, a potential vertical avoidance response, but this was not studied in sperm whales in greater detail.

continued

TABLE 2.1 Continued

Experiment/Observations	Results
Hawaii Results	
Observations from shore stations during ATOC transmissions (Frankel and Clark, 1999a,b)	• Analysis of whales observed near shore, approximately 14 km from the ATOC source, showed a statistically significant increase in the time and distance between successive surfacings as a function of estimated received level of ATOC transmissions. These whales were exposed to levels up to 130 dB in the 60- to 90-Hz band. • Analysis of sighting rates of inshore whales showed slightly higher rates during control versus transmission periods, but the difference was not statistically significant. No difference was found in the distance of whales from shore, but distance from the source was not analyzed.
Observations from boats during playbacks (Frankel and Clark, 1998a)	• Only 11 of 50 playbacks exposed whales to received levels >120 dB in the M-sequence band.[a] These whales did not show a pronounced avoidance response. • Two behavioral variables showed a significant increase with increasing exposure: the distance traveled and time taken between successive surfacings.
Results of bottom-mounted recorders (Frankel and Clark, 1999b)	• No difference in the amount of energy in the band of humpback song was detected comparing 20 minutes before transmission, during, and after transmission, from one recorder placed offshore near the ATOC source and four recorders placed inshore near the main concentration of whales.

ATOC MMRP Conclusion	NRC Committee Conclusion
• The distribution and abundance of inshore whales did not change significantly, but there were significant changes in diving behavior that increased with increasing received level, up to 130 dB.	• There is a data gap in testing for changes in the distribution and abundance of whales near the ATOC source. Given evidence for changes in the diving behavior of whales exposed to low levels far from the source, there is a clear need to study changes in the behavior of whales near the ATOC source in order to evaluate the potential impact of behavioral disruption. It is uncertain whether a change in the time and distance between surfacings is a biologically meaningful measure of the effects of the ATOC source.
• Whale tracks and bearings did not differ significantly between control and playback conditions. • Overall, subtle responses to M-sequence playbacks could only be detected statistically, but the biological significance of these responses is uncertain.	• The limited number of whales exposed to received levels >120 dB limits the power of overall tests. • The same behavioral responses were observed in scaled playbacks as were detected in the operation of ATOC. This replication increases confidence in the robustness of this response.
• There was no change in received song power in response to ATOC. This does not mean that no whales stopped or started singing in response to transmissions but rather that the average song level did not change. There was no widespread change in singing behavior or movement away from the area.	• A comparison of total sound energy in the band of humpback song power shows no change just before, during, or after exposure. This is a very crude response measure, which would miss potentially important responses (e.g., half of singers stop, half double their source level). In addition, much of the power in the "song band" could stem from sources other than humpback songs and this unmeasured background would not be expected to change in response to an ATOC transmission. This dilutes the power of the test. Details of the movement patterns and songs of individual whales singing near the source need to be studied before it will be possible to evaluate the effect of ATOC on singing whales fully. The appropriateness and statistical power of this method were not tested.

continued

TABLE 2.1 Continued

Experiment/Observations	Results
Results of aerial surveys (Mobley et al., 1999)	• Of five major areas in the Hawaiian Islands, Kauai/Niihau had the second-highest number of humpbacks sighted, after the four-island region west of Maui, but Penguin Banks west of Molokai had the highest density after corrections for observing effort. • An average of 5.3 humpbacks and 0.6 sperm whales were sighted per survey during 1994 within 40 km of the source location before ATOC transmission. An average of 7.0 humpbacks and 0.75 sperm whales were sighted per survey within 40 km of the source during 1998 after the source was activated. This higher rate in 1998 may reflect better sighting conditions. • Humpbacks prefer water <200 m deep, but 30 percent were sighted in 200-2,000 m depths. • Mean distances from shore and from the source were higher for humpback whales when the source was on, but difference was not statistically significant.

NOTE: The first two columns of the table were provided to Christopher Clark and Dan Costa for review before the report was published.

aAn M-sequence signal is a phase-modulated tonal signal.

compared with after at least 4 days without ATOC transmissions. Acoustic surveys also were planned by MMRP investigators, and acoustic measurements were attempted using ATOC receivers. The planned acoustic surveys did not yield any data because of the failure of deployed equipment.

The following null hypotheses were tested using aerial, visual, and acoustic surveys (ARPA, 1995, p. C-12):

H_0: There is no detectable difference in sighting rate, distribution, orientation, general activity, or group size of different species (or groups of species) between

ATOC MMRP Conclusion	NRC Committee Conclusion
• For humpbacks no significant changes in distance from source or distance from shore were noted from the 1994 results, when the ATOC source was not operating, to 1998, when it was operating. • Sperm whales, previously described as infrequent in Hawaiian waters, were found offshore of all five island regions.	• Humpback whales in Hawaii showed a pattern of increased distance from the source when it was operating (average = 19 km) compared to off (average = 17 km). This difference, based on 28 exposure sightings, was not statistically significant. A similar pattern in California (16.9 km for exposure versus 14.5 km for control), based on 105 exposure sightings in the 40 * 40 km inner box, was highly statistically significant. It is likely that humpbacks in both places have similar responses, and that difference in significance stems from limited sample sizes in Hawaii. The possible shift in distribution at great ranges from the source suggests the need for a behavioral study on responses to ATOC signals targeting animals near the source. • Kauai is an important habitat for the expanding population of humpback whales wintering in Hawaii. Coarse analysis does not suggest a mass evacuation of the area within 40 km of the ATOC source. However, enough humpbacks were sighted in water >200 m depth to justify a targeted study of humpbacks near the Kauai source location. The presence of offshore sperm whales also suggests that a targeted study on the impacts of long-term ATOC signals on this endangered species would be appropriate.

surveys conducted when the source is on and when it is off, and as a function of distance from the source.

H_0: There is no detectable difference in cetacean acoustic behavior (i.e., call types, rates, structure, or sequence patterns) between measurements from recordings made when the source is on and when it is off, and as a function of distance from the source.

Appropriate tests of these hypotheses assume that the precision of the measurements and the statistical power of the tests are great enough to demonstrate actual differences; that is, the probability of a false negative result is small. Studies of marine mammals in the wild are so difficult (due to problems of observing animals that spend much of their time underwater) and the results so imprecise (because of natural variability and low sample sizes) that it is easy to imagine that such studies might not detect differences that could reflect biologically significant impacts. Only for tagged elephant seals exposed to California transmissions were analyses of precision or power presented, and thresholds for biological significance were not suggested. These factors make it difficult to evaluate the validity of the MMRP's negative results, especially for species other than elephant seals.

Aerial surveys were conducted from November 1995 to October 1998. During control surveys there were 1,524 marine mammal sightings[4] involving 29,826 animals. During experimental surveys (source on), there were 1,617 marine mammal sightings, involving 27,874 animals. Not only were there more sightings in both the experimental and the control conditions than expected, there was a larger diversity of marine mammals sighted than expected. At least eight species of small- and medium-sized toothed whales were observed, four species of seals, five baleen whale species, and two species of large toothed whales. The most frequently sighted large whales were humpback whales (482 sightings) and sperm whales (349 sightings), numbers large enough to permit statistical tests for differences between control and experimental surveys. Statistical analyses of these data were not completed by the time of the Committee's April 1999 meeting. Aerial surveys revealed that humpback whales were significantly further from the source when it was on than when it was off. A similar pattern was found for sperm whales, but the statistical significance was complicated by seasonal differences in distribution (Calambokidis, 1999). Calambokidis also found an increasing trend in the number of humpback whales off the U.S. west coast from 1988 to 1996 (6.7 percent) and from 1996 to 1998 (9.3 percent), using photoidentified whales and mark-recapture calculations, indicating that the ATOC source did not negatively affect the population level of this species.

Elephant seals are important research subjects in relation to the effects of the ATOC source because they have sensitive low-frequency hearing (Kastak and Schusterman, 1998), swim in the pelagic zone, and routinely dive near the depth of the deep sound channel.[5] This species has breeding areas near the California ATOC source site, and these animals are excellent subjects for tag attachment (tags are subsequently removed or shed during molting; D. Costa, University of

---

[4]A sighting is one group of marine mammals, regardless of number.

[5]The deep sound channel or SOFAR (SOund Fixing And Ranging) channel occurs at a depth in the ocean at which "sound rays propagating close to horizontally are trapped by refraction, reducing spreading loss and avoiding surface and bottom losses" (Richardson et al., 1995). The SOFAR channel is found at approximately 1,000 m in the open ocean and approaches the sea surface in polar regions.

California, Santa Cruz, personal communication, 1999). Satellite tags were used to track the locations of naturally migrating individuals. A total of 26 adult males were followed during their natural migration; five tracks were observed when the source was on, and 11 tracks were monitored when the source was off. Only a few tracks of these naturally migrating seals passed near the ATOC source, and there was no obvious avoidance, based on nearest approach to the ATOC source. Based on aerial surveys, elephant seals were found at the same distance from the source whether the source was on (50 sightings) or off (35 sightings).

Translocation experiments were used to obtain a larger sample size of seals exposed near the source. In these experiments, archival tags designed to record received levels of sound and dive patterns were attached to juvenile elephant seals removed from a rookery. Thirteen elephant seals were translocated near the ATOC source when the source was operating, and five seals were translocated to the same site when the source was off. The maximum measured received levels of the ATOC sound for each of the 13 seals in this experiment ranged from 118 to 137 dB. MMRP investigators conducted an extensive statistical analysis of dive patterns of the translocated seals (including a variety of measures, such as time of dive and depth) comparing (1) the dive before the source was turned on, (2) the first dive that started after the source was turned on, (3) the second dive after the source was turned off, and (4) an average of dives measured over 18 hours following the second dive after the source was turned off. The comparisons conducted thus far suggest that there was not a statistically significant change in the dive behavior of translocated seals in response to ATOC transmissions at received levels of 118 to 137 dB. However, the preliminary statistical analysis comprised hundreds of individual tests. These must be merged into one overall analysis, with proper correction for significant results that can occur by chance when a large number of statistical tests are run.

Hawaii ATOC Source

The Hawaiian observations focused on humpback whales and were planned to include aerial visual surveys, passive acoustic monitoring, and shore-based surveys of reactions to ATOC transmissions (off Kauai) and playbacks of hump-back whale vocalizations (off Hawaii). The only peer-reviewed paper analyzing the responses of marine mammals to the ATOC signal to date is that of Frankel and Clark (1998a). This paper did not report on research involving the ATOC source in its site off the north coast of Kauai but described a series of playback experiments using a smaller vessel-deployed source off the coast of Hawaii in a much better site for observing whales. This source was operated at 172 dB, with a frequency bandwidth of 60 to 90 Hz (the same as ATOC). The source was a vessel moored each day in an area of high whale density off the leeward coast of Hawaii, in a position allowing excellent monitoring of humpback whales from a shore station. Unlike most MMRP observations, timing of operation of the source was determined by whale monitoring rather than the predetermined ATOC transmission schedule. In this study, if the shore observers could follow a whale or group of whales for 25 minutes, they would radio the playback vessel and instruct the boat to start an experiment. On a randomized schedule, during 50 of the 85 trials the vessel transmitted the ATOC signal; the other 35 trials were no-sound controls. The shore observers were unaware of which condition was being employed during any given trial. The shore team attempted to continue to observe the whales during the 25-minute experimental period and for a 25-minute post-trial phase. The estimated received level (based on empirical measurements at different ranges and bearings from the playback vessel) at the whales' location during playback varied from ambient (near 90 dB) to nearly 130 dB.

Statistical analyses of whale tracks and swimming directions revealed no difference in these factors between experimental and control conditions. However, this is difficult to interpret because the analysis combined data from whales located so far from the playback signal that the signal was buried in ambient noise, with only a few whales exposed to received levels high enough to expect the possibility of a response. Simple nonparametric comparisons of speed, duration, and distance between surfacings of the humpback whales also showed no difference between control and playback conditions. It appears that the swimming direction of whales with respect to the playback source was not analyzed, even though this is the critical measure for determining an avoidance response. There was a slight significant increase in the time and distance between successive surfacings at increasing received levels of playback; this strengthens the Committee's concerns about conclusions of no effect in these pooled data. Of the 85 trials, an observed whale passed within the 120-dB isopleth at a range of about 400 m in 11 playback trials and five control trials. A potential avoidance reaction was observed in one of these 11 playback trials; a similar "avoidance reaction" also was observed during control observations with no sound. Two potential approaches were observed during playback. The limited sample size of animals exposed to received levels greater than 120 dB limits the power of conclusions regarding lack of effects.

Although the sample size of whales exposed to playbacks louder than a 120-dB received level was small, the results imply that most whales would be unlikely to show an avoidance response when exposed to sound levels of 90 to 130 dB. The responses observed were no stronger than those elicited when vessels approached whales in the study area.

The Committee received several unpublished manuscripts from MMRP investigators about the responses (or lack thereof) of humpback whales to the ATOC source 14 km north of Kauai's coast. The Committee was told that aerial survey results suggested there may be resident populations of sperm whales and **short-finned pilot whales** (*Globicephala macrorhynchus*) in the offshore waters, but the Committee was not presented with any data on the distribution or potential responses of these two species when exposed to the ATOC sound. Rather,

MMRP studies concentrated on humpback whales, as indicated in the EIS for the Kauai source (ARPA and NOAA, 1995). The observation effort focused on inshore waters roughly 10 km from the source. Transmission loss measurements suggest that the received level of the ATOC signal in the inshore waters preferred by humpback whales (less than 200 m deep) is less than 120 dB in the 60- to 90-Hz frequency range.[6] Although many analyses found no difference in responses between control and transmission conditions, some statistically significant differences apparently were observed, even though most whales appeared to be exposed to levels less than 120 dB. As in the playback experiments off the island of Hawaii, the distance between successive surfacings increased with increasing received level of ATOC transmissions off Kauai during 1998 ($p = 0.0017$). This could have resulted if the whales either were swimming faster or stayed under water longer between surfacings.

The Quick-Look Report of the Hawaii ATOC-MMRP (Frankel and Clark, 1998b) (an unpublished and unreviewed account)[7] concluded that "there were no acute or short-term effects of the ATOC transmissions on marine mammals."[8] The Committee questions whether a conclusion this broad can be reached at this time using the data provided. The report does, in fact, present evidence for some short-term behavioral changes in response to the ATOC sound source by humpback whales. Even more important, the Committee questions the ability of the MMRP to show the absence of any response. The failure to observe an effect could result from a number of factors, including the specific conditions of the experiment and lack of sufficient statistical power (resulting from an insufficient number of observations or the statistical test chosen). This concern is particularly heightened for the Hawaii MMRP study in which most observations were made far from the source and no results were presented on responses of the marine mammal species most commonly sighted offshore near the source (sperm whales and pilot whales).

Contrary to plans listed in the EIS for the Kauai source (ARPA and NOAA,

---

[6]The original predictions in the initial EIS were based on a spherical transmission loss from the 20*log(range) relationship. That is how the 40-km radius "zone of influence" was determined. This contour is approximately 7.5 km south of the ATOC source. Actual measurements of the ATOC transmissions found that the 120-dB received level occurred approximately 4.8 km south of the source. At the 200-m contour, the received level was approximately 111 dB in the 60- to 90-Hz ATOC band (C. Clark, Cornell University, personal communication, 1999). This was more consistent with the predictions of a cylindrical equation model, which terminated the 120-dB isopleth at the 200-m depth contour.

[7]The Quick-Look reports were undoubtedly designed as a means to disseminate research results rapidly and widely to try to achieve open access to research results and keep the public informed, both worthy goals. However, the Committee determined that the Quick-Look format was generally counter-productive because it widely disseminated non-peer-reviewed results and did not encourage timely peer review and publication of research results.

[8]http://atoc.ucsd.edu/HIquicklookrpt.html, accessed 10/13/99.

1995), there were no analyses of the vocal behavior of individual humpback whales exposed to the ATOC signal. Instead, there was only a general assessment of the total energy at humpback whale vocalization frequencies in the area, which could be misleading given that all vocal activity in the area was summed for pre- and postexposure. Considering the analyses conducted to date, the possibility that the ATOC signal might affect the song production of humpback whales cannot be eliminated. We do not know the function(s) of humpback songs, but they may be a reproductive "advertisement" display, as for the songs of some birds (Tyack, 1981).

The limited data presented by the MMRP made it impossible to draw any but the most tentative conclusions about the effects of ATOC sounds. Based on the material presented, baleen whales, sperm whales, and elephant seals in California, and humpback whales in Hawaii did not show profound avoidance responses to the ATOC signal. However, complete analyses and peer review are required before any more definitive conclusions can be reached.

COMPARISON OF THE RECOMMENDATIONS OF NRC (1996) AND MMRP RESPONSES

Although it was difficult to evaluate the MMRP in midcourse, the 1996 NRC interim report contained the following conclusions:

1. The California ATOC source transmissions did not appear to cause a major change in the distribution of marine mammals.

2. The constrained sound characteristics and conditions used during the MMRP-controlled ATOC transmissions impeded the project's ability to answer fundamental questions concerning the impact of ATOC-like noises on marine mammals.

3. Several changes in the plans for the Hawaiian MMRP studies (eventually concluded in 1998) could provide more definitive information about the potential of ATOC sound to affect marine mammals and other organisms. Specifically, shore-based observations should be conducted for the entire 6 months of ATOC transmissions, and the effectiveness of observational methods should be validated using playbacks of relevant natural sounds conducted within visual range of the shore station.

The Committee reviewed plans for MMRP research in its 1996 interim report and believed that the value of the work could have been enhanced considerably with some modifications in the proposed study. Not only would these changes have strengthened the ability of the MMRP to test the effects of the ATOC sound on marine organisms, the additional data obtained would provide useful insight into broader questions about the effects of low-frequency sounds on marine mammal behavior. Much of the MMRP research focused on statistical tests of whether behavioral indicators varied significantly between transmission and control conditions. The Committee favored tests in which the biological significance of any such changes could be evaluated, which would require a broader investigation into the effects of noise on the normal behavioral ecology of each species.

The goal of the MMRP was to determine whether the ATOC transmissions might adversely affect reproduction or survival of marine mammal populations. The methods available to measure population trends are crude, and the cause of conserving populations will not be met by waiting until a threat actually causes a measurable decline in populations. Therefore, it is useful to use more short-term measures as indicators of potential adverse impact. For example, if marine animals avoid or leave critical habitats because of a human disturbance, the animals may enter suboptimal habitats, with eventual negative effects on feeding and/or reproduction. Proxies selected to measure adverse impact should be easily measured animal behaviors that, if disrupted, would have significant negative impacts on marine mammal reproduction and longevity. The apparent avoidance reactions observed in the California MMRP studies are good examples of relevant measures; the impact can be related to the percentage of habitat lost or can be estimated by comparing the quality of the habitat the whales left compared to that to which they moved.

Other elements of the MMRP studied behavioral changes that are less relevant. For example, the Hawaii MMRP analyzed the distance traveled and time spent between surfacings for humpback whales and found a statistically significant trend for these measures to increase with increasing exposure to ATOC transmissions. Even though these results are statistically significant, it is difficult to interpret their possible biological significance. We suggest that future studies of this sort carefully select behavioral and physiological measures that can more easily be related to potential adverse impact. Basic research in the behavioral ecology of many species is required to direct these choices. For example, the more we know about the foraging ecology of a species, the better we can interpret the biological significance of disruption of feeding behavior, or movement to different feeding areas. Since humpback song is known to be involved in the breeding behavior of humpback whales and the ATOC sound could have disrupted singing, the Hawaii MMRP observations would have benefited from selecting behavioral studies that could more easily be related to potential impact on song and thus breeding behavior.

NRC (1996) offered two suggestions and one point for further consideration. The first suggestion concerned maintenance of the onshore observation station on Kauai for the full duration of source testing to allow time for additional playback experiments. The second suggestion concerned the need for prompt analysis and dissemination of MMRP results. An additional point concerned other marine species that are potentially ensonified by ATOC sounds. For each of these issues the following sections will present recommendations of the NRC from its 1996 interim report, followed by the MMRP responses.

Maintenance of the Shore Station and Playback Experiments

At its October 1996 meeting the Committee was presented with plans for maintenance of the shore observation station at the Hawaii field site during the 1997 ATOC transmissions to observe behavioral changes in humpback whales during exposure to ATOC sounds. Observation of marine mammals from similar shore stations provided useful baseline data during a previous playback study of humpback whales off the coast of the island of Hawaii (ARPA and NOAA, 1995, Appendix G) during the 1993 to 1994 season and earlier off the coast of California (Malme et al., 1983, 1984). According to plans for the ATOC sound transmissions in 1997, the shore station observers were to be in place for only 4 to 6 weeks. This period was designed to provide the minimum amount of information needed for comparison with the 1993 to 1994 baseline data, with no margin for unforeseen circumstances.

The Committee disagreed with this minimal effort and recommended that the shore station be maintained and used throughout the humpback whale season off Kauai (e.g., during the entire 6 months of MMRP-controlled ATOC transmissions). This suggestion was based on the conclusion that additional very useful data (see below) could be obtained from continuation of the shore-based observations, especially with the addition of natural sound playbacks near the shore station. Extending the field season also would have increased the sample size of observations, making it more likely that significant effects of the ATOC transmissions would have been detected, if such effects actually occurred. Shore-based observations are important because they provide a means of observing marine mammals without introducing the confounding effects of nearby vessels. Although the shore site was probably outside the area within which an effect would be expected, such observations should have been able to determine whether inshore humpback whales, rather than offshore near the ATOC source, would be affected. The MMRP did conduct shore observations for 6 weeks from February 9 to March 20, 1998. Observations were not extended beyond this time.

According to the MMRP, as of 1996, ATOC transmissions in California and an ATOC-like test signal played off Hawaii had little observed effect on marine mammal behavior, at least in terms of surface tracks and the number, frequency, or depth of dives. Interpretation of these findings is complicated, however, because there had been no observed response to ATOC signals. Thus, it was impossible to establish the validity of the method that had been used to study ATOC's effects. Simply stated, the Committee could not choose between the conclusion that the ATOC signal had little or no effect and the alternative view that the observational methods used were not sensitive enough, or not designed appropriately, to detect the effects of these sounds. In the interest of facilitating future investigations into the effects of ATOC or ATOC-like sound sources on marine mammals, it is essential that an effort be made to define protocols that are useful scientifically and relevant to the decisions that must be made. Such protocols could specify how observers must confirm before their observation program begins that their observation techniques can measure the variables of interest—behavior associated with critical activity—as well as the minimum statistical power that can be tolerated, so that significance of disruption can be ascertained. Analyses should be framed not only to test for any detectable change, but also to estimate the percentage of time a behavior is disrupted, the amount of energy wasted, and/or the probability that the disruption will prevent animals from achieving the goal of the activity.

The Committee's dilemma could have been overcome if the MMRP had been able to demonstrate that its observation methods were valid. The most direct means to test the methods would have been to increase the intensity of the sound source until a response was observed, to create a direct estimation of the stimulus-to-response relationship. However, the ATOC source cannot be operated at levels higher than 195 dB for technical reasons, and it is unknown whether higher levels would produce a measurable response without being unduly harmful to marine mammals. Therefore, the Committee suggested an alternative approach—that of broadcasting noises other than ATOC signals that would affect marine mammal behavior in a way that is detectable by the same (or similar) methods used in the ATOC study. In earlier studies of marine mammal responses to playback of auditory stimuli, Clark and others (Clark and Clark, 1980; Malme et al., 1983; Tyack, 1983; Mobley et al., 1988; Frankel et al., 1995) have shown that animals respond strongly to certain natural vocalizations, such as the calls given by other members of the same species or the vocalizations of a predator, such as the **killer whale** (*Orcinus orca*).

Although use of a non-ATOC stimulus would not allow validation of the specific response to the ATOC stimulus, it would at least have validated that MMRP investigators could observe (with sufficient precision and accuracy) such things as startle, flight, and vocal responses, which could be produced by many different stimuli, whether ATOC sounds or killer whale calls. The Committee recommended that the MMRP incorporate natural sounds into its research during its 1997 to 1998 studies, taking into account the results of the playback studies cited above. The MMRP did not conduct extensive vessel-based playbacks of natural sounds, although a vessel was used during this time to observe whales. Frankel and Clark (1998a) reported on one playback of an Alaskan humpback whale feeding call, although the results were ambiguous. If playback of these natural sounds had elicited a strong observable behavioral response from the subjects, that response would have provided an important validation for the observational method used by the MMRP to test ATOC's effects on the behavior of marine mammals. Measurement techniques must be calibrated with a stimulus that produces a measurable response. Such a calibration allows a scientist to distinguish between a true lack of response and a response that was unmeasurable by the chosen technique. For example, in the case of singing humpback whales, had the researchers tested enough singing animals with a biological sound (e.g., killer whale vocalizations), they could have determined which behavioral parameters showed statistically significant changes (e.g., song elements, diving behavior, evacuation of the area) and provided a baseline of comparison for other stimuli. Because the MMRP did not add this component, the difficulty of interpreting the MMRP's results remains.

Changes that might have indicated significant effects of ATOC transmissions include

- significant changes in singing patterns (would need to correlate with calving rates);
- significant flight of animals from the source area (significant either in distance, speed, duration, or movement into harm's way);
- significant reduction in calves produced; and/or
- significant abandonment of area by identified individuals in later years.

Need for Prompt Data Analysis

One of the problems faced in preparation of the 1996 NRC report was the lack of analyzed data from a number of MMRP field studies, particularly those conducted in Hawaii. Thus, it was difficult to assess the quality and significance of this work and to make suggestions for future ATOC-related marine mammal studies. In its 1996 report the NRC noted that such a situation, if it persisted, would compromise the Committee's future work, and it would not be able to conclude whether there are deleterious effects of the ATOC sound source on marine mammals (or other organisms). The Committee expressed its hope that it would receive a full analysis of MMRP observations and conclusions at least 2 months prior to its final meeting. The NRC strongly recommended that data analysis and presentation be the highest priority for investigators in both Hawaii and California and that sufficient funds be set aside to enable a complete and expeditious evaluation of the data. The Committee's 1999 meeting was scheduled for approximately 6 months after completion of MMRP observations to allow time for analysis to be completed. The Committee received the June 1998 Quick-Look Report on the Hawaii ATOC-MMRP Results several weeks before its April 1999 meeting. This Quick-Look Report included some preliminary analyses and indicated that more extensive analyses would be forthcoming. The Committee did not receive any completed analyses or conclusions before the meeting. The MMRP investigators at the meeting asserted (as did the Quick-Look Report) that considerable further analysis was needed to interpret the data properly. In discussions with MMRP investigators during the April 1999 meeting, it was clear that only limited funds and personnel were available during the final year of the MMRP and that this shortage continues to jeopardize the quality and timeliness of the scientific products of the MMRP.

Ensonified Species Other than Marine Mammals

In addition to the two specific suggestions for MMRP research, the Committee noted that the EIS for both the Hawaii and the California sources included analyses of the effects of the ATOC sound on other biota, including marine turtles, fish, and other organisms (ARPA, 1995; ARPA and NOAA, 1995). The only study published for other vertebrates from ATOC-funded research was for an experimental study of ATOC-like sounds on fish (Klimley and Beavers, 1998).

The NRC (1994, pp. 53-53) specifically pointed out that a major concern for all low-frequency ensonification is not only effects on marine mammals but also the potential effects of such sounds on other components of marine mammal food chains, such as fish or zooplankton, and on other endangered species (e.g., turtles). The Committee strongly supports this assertion and continues to be concerned that low-frequency sound may have implications for a far wider range of the marine biota than is being studied at the moment. This is of particular importance for sound sources such as ATOC that will be operated in one place for years at a time. In addition, a number of studies suggest that ATOC-like sounds may be very attractive to many species of sharks (Myrberg, 1972, 1978; Myrberg et al., 1976). Sharks attracted to ATOC sounds could be affected adversely by these sounds, and ATOC equipment could be jeopardized by sharks. The lack of study on marine organisms other than mammals makes it impossible to infer the potential impact of long-term deployment of ATOC-like sources in areas used by sensitive species.

SIGNIFICANCE OF THE MMRP TO RESEARCH USES OF SOUND

Data presented by the MMRP were inconclusive regarding the effects of the ATOC sound on marine mammals. The Committee considers that existing data from the MMRP and other sources—such as recent work motivated by the 1994 NRC report and funded by ONR (e.g., Au et al., 1997)—suggest, however, that there is *no cause for alarm* about the short-term effects of ATOC sources on dolphins and most seals because they do not dive to depths that would allow them to encounter the source at levels they could hear well. However, there *is cause for concern* because we cannot totally rule out short- and long-term effects of ATOC, particularly on baleen whales and sperm whales. Optimally designed studies are needed on the long-term effects of high-intensity sound sources (e.g., interference with communication and reproductive activities, exclusion from critical habitat).

ATOC investigators plan to apply for funding and permission to continue ATOC transmissions in Hawaii for another 5 years (the California source has been terminated). ATOC investigators plan to conduct aerial surveys near the Kauai ATOC source to monitor the distribution and abundance of marine mammals to advance the understanding of possible long-term acoustic impacts (P.

Worcester, Scripps Institution of Oceanography, personal communication, 1999). It is outside the Committee's charge to comment on whether ATOC should be allowed to proceed. However, if it does proceed, monitoring of marine mammal behavior and responses to the ATOC transmissions should continue as an integral part of the experimental design in order to improve the ability to evaluate the impact of ATOC during the next 5 years of Hawaii ATOC transmissions. As part of this continued evaluation, there should be annual reports of all yearly data to an oversight body not associated with ATOC (e.g., the Marine Mammal Commission or National Oceanic and Atmospheric Administration), with the authority to terminate transmissions if there is evidence of significant deleterious effects from long-term exposure. ATOC scientists should be required to notify the oversight group immediately if they detect significant adverse effects on marine mammals. Continuation of ATOC transmissions should be conditional on timely publication of marine mammal results. Chapter 5 includes a discussion of appropriate monitoring that should be considered if ATOC is approved to continue.

Assessment of
Continuing Research Needs

In its 1994 report the NRC recommended future research that would provide a better understanding of the effects of low-frequency sounds on marine mammals and their prey. This chapter assesses the progress made in addressing the targeted issues since publication of that report. The major aims are presented as described in that 1994 report; specific goals are described in boldface type.

Since 1994 the Office of Naval Research (ONR) has invested significant funds into attempts to address a number of the issues raised in the 1994 report. These studies were described to the Committee by Robert Gisiner at the April 1999 meeting, and their results are cited below as appropriate. Although an excellent start has been made in addressing the many questions, there is still a dearth of data on marine mammal bioacoustics. Some of these research needs fall within the purview of ONR, but it is a mission agency and its goals are highly oriented toward the missions of the Navy. Thus, it cannot be expected to deal with all of the important issues raised in the 1994 NRC report, implying that additional sources of funding for marine mammal bioacoustic research are required if better knowledge of marine mammal hearing is deemed by policy makers to be desirable.

Since publication of the 1994 NRC report, the anatomy of the inner ear of several additional marine mammal species has been studied. Computational models based on the anatomical parameters of marine mammal cochleas have been developed, and predictions from such models have correlated well with behaviorally determined audiograms in several species of toothed whales (Ketten, 1997). This kind of modeling provides an important new tool for assessing the auditory sensitivity and frequency range in whales that are not amenable to experimental measurements. There have been significant advances in our knowledge of low-frequency hearing capabilities of several toothed whale species as well as in techniques required to acquire such data. Significant data have been obtained on temporary threshold shift (TTS)[1] in several marine mammal species (Ridgway et al., 1997; Kastak and Schusterman, 1998; Kastak et al., 1999; Schlundt et al., 2000). Additional work needed includes (1) anatomical studies of species with known audiograms to validate the use of anatomical features to predict auditory capabilities and (2) studies using both behavioral and auditory evoked potential[2] techniques to determine auditory capabilities of marine mammals (especially of baleen whales). Additional studies are needed to determine the abilities of marine mammals to detect natural sound in the presence of human-generated background noise (Erbe and Farmer, 1998).

Several papers have appeared since 1994 reporting on use of auditory evoked potentials to study dolphins (Szymanski et al., 1995, 1998; Popov and Klishin, 1998; Popov and Supin, 1998; Popov et al., 1998). Dolphin (1997) reviewed auditory processing in several toothed whales (*Grampus, Orcinus, Tursiops, Delphinapterus*) that were studied using evoked potential methods. He pointed out that auditory evoked potentials can be used to study a wide range of questions about hearing and the auditory system. These include determination of auditory filter shapes that may provide clues about the potential masking effects of some human-generated sounds, determination of TTS resulting from sound exposure, and studies of the masking effects of specific sounds. Dolphin (1995, 1996) and Dolphin et al. (1995) examined temporal processing in several whale species in response to amplitude-modulated stimuli using evoked potential techniques. Ridgway and Au (1999) reviewed earlier work on processing by the auditory central nervous system and approaches to sound conduction to the dolphin ear. Popov and Supin (1998) studied dolphin auditory evoked responses to rhythmic sound pulses. Popov and Klishin (1998) reported on a study of common dolphin hearing using the electroencephalogram. Popov et al. (1998) reported on frequency tuning of the dolphin auditory system using evoked potential methods.

---

[1]Temporary threshold shift is a temporary increase in the threshold audible sound level presumed to be caused by temporary inactivation of the outer hair cells at a given frequency.

[2]Auditory evoked potentials (AEPs) are electrophysiological recordings of minute voltages generated by neural activity in the brain in response to acoustic stimuli. AEPs can be noninvasively recorded from the scalp skin surface and have been broadly applied with great success in humans. The response following the presentation of a brief acoustic stimulus is a series of peaks or waves that arise as a consequence of more or less simultaneous firings in sets of neurons located in successively higher auditory nuclei. Evoked potentials are quite weak, meaning that multiple presentations of the acoustic stimulus and averaging techniques must be used to measure them. Among the advantages of AEP measurements are that (1) they require no or only minimal cooperation from the subject, (2) responses are rapidly obtained and are highly robust, (3) response detection can be fully automated and based on totally objective acceptance criteria, and (4) tests are noninvasive and therefore amenable to examination of protected species. AEPs have now been obtained from a wide range of species. The availability of such data greatly facilitates comparative studies of hearing and auditory function.

Employing the auditory brainstem response (ABR), a type of evoked potential technique, Szymanski et al. (1999) reported that the ABR of killer whales provided a means of suprathreshold hearing measurement. The ABR thresholds of individual killer whales were, at most frequencies, within 12 dB of thresholds measurable by behavioral responses. At the most sensitive frequency (20 kHz), mean thresholds determined by behavioral and physiological methods differed by only 3 dB, thus showing the usefulness of this technique as a proxy for measurements of behavioral thresholds. The two killer whales studied are the largest animals (2,000 to 3,000 kg) ever successfully investigated with evoked potential methods (see also Szymanski et al., 1995, 1998). Although the ABRs were lower in amplitude than those for dolphins less than one-tenth as heavy, ABRs from the killer whale (Szymanski et al., 1999) were adequate for determining values near the behavioral threshold (ABRs averaged 5 dB higher than behavioral thresholds for the most sensitive range of 18 to 42 kHz). Because ABR amplitudes appear to correlate with the relative size of auditory structures in the brain, Szymanski et al. (1999) suggested that successful use of this method in the larger baleen whales may be more difficult. Experience with a young gray whale at Sea World (San Diego) demonstrated that working with larger whales requires more sensitive techniques, quieter conditions, and more time to use ABR techniques on whales than is needed for smaller marine mammals. However, the promise of this technique for validating behavioral observations on large whales indicates that the NRC (1994) recommendation about testing hearing in beached, stranded, or entrapped larger whales should continue to be a goal. These investigations should build on those mentioned above using teams experienced in electrophysiological techniques.

A comprehensive set of species groups, signal types, and biological parameters that should be measured for marine mammals is presented in Chapter 5. Priority for acoustic studies should also be given to species that are (1) endangered or threatened (e.g., the **northern right whale**, *Eubalaena glacialis*); and (2) known or suspected to hear and communicate using low-frequency sound (e.g., baleen whales, sperm whales, elephant seals).

BEHAVIOR OF MARINE MAMMALS IN THE WILD

Aim: To determine the normal behaviors of marine mammals in the wild and their behavioral responses to human-generated acoustic signals (NRC, 1994, pp. 41-47). This aim can be conceptualized as a number of specific topics, which are shown here in bold print.

- **Determine how marine mammals utilize natural sound for communication and for maintaining their normal behavioral repertoire(s).**

Although much is unknown about communication among baleen whales, and little research is currently directed at this topic, some work has been published since 1994 on species that may be affected by the Acoustic Thermometry of Ocean Climate (ATOC) signal or low-frequency active (LFA) sonar. **Gray whale** (*Eschrichtius robustus*) vocalizations were documented at two locations along their southward migration by Crane and Lashkari (1996), who found that vocal behavior varied with location. More vocalizations occurred in shallow water than in deep water, and the type of vocalization (call structure) produced most frequently during the southward migration differed from vocalizations in the lagoons of Baja California where the whales calve and spend the winter. Crane and Lashkari concluded that vocal activity is an important component of migratory behavior in gray whales, probably for communication rather than navigation. These data are important as a reminder that the potential for acoustic interference by human-generated noise can be site-dependent and seasonally variable. Clearly, there is a greater potential for disruption of normal behavior in areas where vocalization rates are high.

Recordings of blue whale vocalizations in the Pacific Ocean have indicated that the occurrence of a two-part call is characteristic of populations in both the Gulf of California and the waters off California (Thompson et al., 1996; Rivers, 1997; Stafford et al., 1999). Whether these two populations are part of a single Pacific Ocean population or whether there is limited exchange is yet to be determined. The presence of a subspecies (**pygmy blue whale**, *Balaenoptera musculus brevicauda*) off Australia (Ljungblad et al., 1997) adds further incentive for documenting the occurrence and types of vocalizations of blue whales in the Pacific Ocean (Stafford et al., 1998).

The recent availability of the Navy's Integrated Undersea Surveillance System (IUSS) and other hydrophone arrays to scientific researchers has permitted the use of passive acoustic tracking as a method for documenting migration routes and critical habitats where baleen whales are seasonal visitors or residents (e.g., Stafford and Fox, 1996; Stafford et al., 1998, 1999; Watkins et al., 2000). Although direct visual observations of vocalizing whales provide more details of behavior compared to the use of remote monitoring (where animals are not observed as they produce sounds), the detection of well-described species-specific calls to track baleen whale migrations and activity using the IUSS and other hydrophone arrays has great potential.

Several studies (listed above) have provided descriptions of vocal behavior, and such data are critical for interpreting the biological significance of any changes in vocal behavior induced by human-generated noise. A shift in the focus of regulations from harassment to the biological significance of behavioral disruption (recommended later in this report) will require a much better understanding of the functions of vocal behavior. Therefore, a high priority should be given to basic research on the behavioral ecology of how marine mammals use the sounds they produce; the results of such research would have immediate regulatory implications.

- **Determine the responses of free-ranging marine mammals to human-generated acoustic stimuli, including repeated exposure of the same individuals. How is the use of natural sounds altered by the presence of human-generated sounds?**

ATOC transmissions from a fixed source off Hawaii theoretically could provide some data on repeated exposure to this signal; however, few individuals were seen close enough to the Hawaii source site either in experimental (ATOC signal on) or control (ATOC signal off) conditions to provide evidence of any response to source levels greater than 130 dB. Individual identification of humpback whales was part of the MMRP, but the activities of specific individuals exposed to the ATOC stimulus repeatedly have not been presented to date. In addition, humpback whale song was recorded prior to and during exposure to the ATOC signal, but these data also have not been examined in sufficient detail to determine if the presence of the ATOC sound has a concurrent and/or long-lasting effect on the songs of individual whales. The Quick-Look reports did indicate that there was no change in average acoustic energy in the 200- to 800-Hz band, a dominant frequency band for humpback song. Detailed examination of the actual songs for changes in frequency content or amplitude of individual components that may overlap with the broadband components of the ATOC signal was not completed at the time of the Committee's April 1999 meeting. Clearly, such data are unique and warrant a detailed examination. If subsequent analyses of data support the finding of no major change in the vocalizations of humpback whales, that would support the contention that ATOC is unlikely to have long-term effects on this species at this important breeding site.

Two studies on **white** whales (*Delphinaptera leucas*) have provided pertinent data since the publication of NRC (1994). Erbe and Farmer (1998) trained captive animals to detect vocalizations from other white whales in the presence of background noise to assess the masking[3] effects of icebreaker activities. Erbe and Farmer found that the aeration system used to clear ice debris had the greatest masking effect, followed by propeller noise. This innovative protocol may be useful for determining masking effects in other toothed whales using a variety of human-generated noises, including the ATOC signal and LFA sonar. A logistically challenging field study was conducted in the St. Lawrence estuary with free-ranging white whales (Lesage et al., 1999). The vocal activity of the animals was monitored prior to, during, and after exposure to two vessels with different

---

[3]Masking is the reduction in the audibility of one sound due to the presence of a second sound. Of greatest interest here is simultaneous masking, in which both sounds overlap in time, but masking can also occur when the signal is a brief sound presented immediately before or immediately after the masker (backward and forward temporal masking, respectively). In humans the amount of simultaneous spectral masking observed is related to the width of the person's auditory filter located at the signal frequency.

noise signatures: a motorboat and a ferry. More than 70 recording sessions were conducted, but only six met the criteria for a successful experimental sequence with acceptable signal-noise ratio for analysis. Despite the small sample size (*N* = three for each vessel), differences were noted in the vocalizations. Calling rates declined in five of the six sessions as noise levels increased, the occurrence of certain call types increased, repetition rates of certain calls increased when a vessel was within 1 km of the white whale, and the mean frequencies produced were higher, presumably to move the frequency of the call outside the frequency band of the masking noise produced by the vessels.

These two studies provide evidence that specific human-generated noises can affect the vocal activity of white whales in the short term. Although generalizations to other species are not without risk, baleen whales exposed to low-frequency noise (from ATOC, LFA, ships) may respond similarly. In addition, the field study by Lesage et al. (1999) indicates that such data are difficult but not impossible to obtain given a sustained field effort.

Tyack (1998) and Tyack and Clark (1998) concluded that 10 of 17 singing humpback whales exposed to low-frequency sounds from the SURTASS-LFA[4] sonar system stopped singing during playback with a source level that ranged from 155 to 205 dB, resulting in maximum received levels of 120 to 150 dB. Four of these 10 stopped when they were joined by other whales, a behavior that has been observed in previous studies with no human-generated sound (Tyack, 1983), and their responses cannot be definitively associated with the playback. There was no difference in the received levels for the six whales that stopped singing, apparently in response to playback, compared to the seven that did not stop, suggesting individual differences in perception and/or response. In order to evaluate the significance of disruption induced by cessation of singing, it will be necessary to make a decision about what proportion of the population of singers can be disrupted before the disruption will have a population-level effect.

An important task is to determine how different sound types and levels affect migration and other movement patterns of marine mammals. The 1994 NRC report specifically recommended shore-based studies similar to the Malme et al. investigation of migrating gray whales. Tyack and Clark (1998) replicated the Malme et al. (1983) study using a sound projector from the SURTASS-LFA. Two important characteristics of the Tyack and Clark study were that (1) the intensity of the source was adjustable and (2) its location could be changed. The results differed considerably depending on both of those variables. When the source level was 170 dB and was located in the migratory path of the whales, animals deflected around the source by a maximum of only several hundred

---

[4]SURTASS-LFA is the acronym for a low-frequency active sonar system developed by the U.S. Navy to detect submarines. This sonar uses a vertical array of sound projectors deployed from a surface ship to broadcast sounds in the 100- to 500-Hz range. The Navy made this system available for three marine mammal studies conducted over a period of 1 year.

meters. When the source level was increased to 185 dB and was located in the migratory path of the whales, the animals changed course so as to avoid passing within a kilometer of the source on both the onshore and the offshore sides. The evidence suggested that most whales avoided exposure to received levels of 140 dB or more. These findings agreed with earlier findings of Malme et al. (1983, 1984) using noises associated with oil industry activities. This study confirmed that whales change their response as source level is changed, which demonstrates that they are responding to received level, not just distance or sound gradient.

When the source was located on the offshore side of the migratory path, there was little evidence of any diversion in the individual migratory paths for source levels of both 185 and 200 dB (Tyack and Clark, 1998). This finding is especially interesting because calculated received levels were higher in these cases than in the condition producing the strong avoidance effect when the source was located in the migratory path at a source level of 185 dB. For example, at the 200-dB source level for offshore playbacks, the received levels measured at ranges of 2 to 2.5 km were >140 dB. Since the offshore source was placed about 2 km offshore from the inshore source location, this means that during the offshore playbacks most of the whales were exposed to received levels that would almost certainly have elicited an avoidance reaction had the source been placed in the inshore location. Thus, for these whales there clearly was something disturbing about a strange sound source located in the migratory path that was not disturbing when that same source was on the offshore side of the migratory path. This study illustrates that behavioral responses to noise sources may not be solely dependent on the acoustic nature of the noise, but on the location of the noise as well. Apparently, the 120-dB avoidance model, which seemed correct for non-impulse noises in the migratory path, is not valid for offshore sources.

High-frequency pingers and submarine sonar pings are known to affect sperm whale vocalization rates and behavior (Watkins and Schevill, 1975; Watkins et al., 1985, 1993). Low-frequency sound also may affect sperm whales because their wide-band clicks contain energy between 100 and 2,000 Hz (Watkins et al., 1985; Moore et al., 1993), which is suggestive of low-frequency hearing.

Gordon et al. (1996) were funded by the MMRP for two three-week cruises off the Azores to study how sperm whales responded to experimental playbacks of M-code sequences similar to the ATOC stimulus as well as similar stimuli of higher frequency. Most standard playbacks were conducted at 75 Hz (the ATOC frequency), and one each was conducted at 2, 3, and 4 kHz. There were 16 control trials. Gordon et al. observed no significant difference between playback and control groups in blow rates of whales before diving or the relative bearing of sperm whales with respect to the source vessel. Sperm whales produce a regular series of clicks starting about three minutes after the onset of a dive. Gordon et al. (1996) compared six different measures from these clicks and found no difference for the 75-Hz playbacks, although the initial click rate was higher for the three higher frequency playbacks than during controls.

Many earlier reports suggest that sperm whales may silence or move out of an area in response to manmade noise (Watkins et al., 1985; Bowles et al., 1994; Mate et al., 1994). The contrasting lack of response in Gordon et al. (1996) may reflect different responsiveness to different stimuli or perhaps that different groups of sperm whales have differing responsiveness depending on their prior exposure history. As with the Frankel et al. (1995) study, this six-week pilot study confirmed the utility of observations for which the study site and experimental protocol are optimized for marine mammal studies. One of the most important limitations of the Gordon et al. study was that the limited source level of the sound source meant that few whales were exposed to received levels above 120 dB. Gordon et al. (1996) advocate further studies using more powerful sources and more sensitive methods for measuring or estimating received levels and for monitoring responses of the whales.

In another recent study of sperm whales in the Atlantic Ocean, Andre et al. (1997) presented individuals off the Canary Islands with various noise sources to determine if acoustic deterrence could reduce whale collisions with ferry boats. Although there is no documentation of the received levels for the sound stimuli used, Andre et al. observed approaches to the source, which were interpreted as curiosity, when an artificial click "coda" was presented. The authors concluded that sperm whales exposed to high levels of shipping noise have a high tolerance for noise. Alternatively, those animals may have permanent threshold shifts. There is no way to distinguish between those possibilities with the data available at present, although postmortem examinations on the cochleas of a few animals could resolve this uncertainty. A careful study of the response of sperm whales to low-frequency sound seems warranted.

Several papers have suggested that beaked whales tend to strand when there are naval operations offshore. Simmonds and Lopez-Jurado (1991) reported on four mass strandings between 1985-1989 of **Cuvier's beaked whale** (*Ziphius cavirostris*) on the coast of Fuerteventura in the Canary Islands that may have been related to naval maneuvers. Frantzis (1998) reported on another mass stranding of 12 or more beaked whales sighted over 38 km of coastline during two days (May 12 and 13, 1996) in the Kyparissiakos Gulf in Greece. There was no external sign of injury or disease in any of these animals. Frantzis (1998) concluded that the mass stranding was associated with a concurrent NATO sonar exercise. The Frantzis paper stimulated the NATO research center that conducted the sonar tests to convene panels to review the data (D'Amico, 1998). The NATO sonar transmitted two simultaneous signals, one at 450-700 Hz and one at 2.8-3.3 kHz at source levels of just under 230 dB. This combined signal lasted four seconds and was repeated once every minute. The NATO analysis suggested close timing between the onset of sonar transmissions and the first strandings. Unfortunately, it was not possible to determine the received levels experienced by the stranded whales. D'Amico (1998) states that received levels as high as 150-160 dB were estimated to occur at ranges of 50 km. Sperm whales were heard within 10-25 km of the sound source, but demonstrated no obvious changes in their clicking patterns before, during, and after sonar transmissions. Although these papers raise concern about the effects of noise on beaked whales, they provide little guidance regarding what exposures may be dangerous and which are safe. There is a clear need for experimental studies of the responses of beaked whales to carefully controlled exposures of noise.

NATO sonars have been tested in the Mediterranean Sea on many occasions without strandings. Both Simmonds and Lopez-Jurado (1991) and Frantzis (1998) started with rare strandings and then looked for some other rare event that might correlate, but neither paper makes a strong case for having performed a thorough systematic survey of when naval or sonar exercises might have occurred in these areas in the absence of strandings. There is a clear need for studies designed to test this association more systematically. In areas where beaked whales are common or have historically stranded, it would be good to set up a prospective study monitoring noise exposure, systematically logging mass strandings and sources of loud noise such as naval exercises. Careful necropsy of stranded animals would help test for any noise-induced injuries.

- **Determine the response of deep-diving marine mammals to low-frequency sounds whose characteristics (source level, frequency bandwidth, duty cycle) duplicate or approximate those produced by acoustic oceanographers.**

The only research on this issue known to the Committee was the previously mentioned study on elephant seal deep diving conducted as part of the California ATOC observations. This should be an area of priority for future studies since it is directly related to the issue of the effects of the ATOC source on the animals that may approach closest to the source.

Several other studies have reported on the effects of ATOC-like sounds on various marine mammals (e.g., Mattlin, 1995; Aburto et al., 1997; Harvey and Eguchi, 1997). However, none of these studies has appeared in the peer-reviewed literature, and weaknesses in data acquisition, analysis, or interpretation limit their usefulness. Although these studies were examined by the Committee, their results will not be considered further. Relevant results should be published in the peer-reviewed scientific literature, and funding agencies should provide support for this critical component of research.

STRUCTURE AND FUNCTION OF MARINE MAMMAL AUDITORY SYSTEMS

Aim: To determine the structure and capabilities of the auditory system in marine mammals (NRC, 1994, pp. 47-53).

This aim can also be conceptualized as a number of specific topics, which are shown below in bold print.

- **Determine basic hearing capabilities of various species of marine mammals.**

Low-frequency audiograms have now been obtained from multiple individuals of several species of toothed whales and seals by using a variety of behavioral and electrophysiological techniques. Audiograms extending below 100 Hz have been obtained for white whales, the **bottlenose dolphin** (*Tursiops truncatus*), the **false killer whale** (*Pseudorca crassidens*), and **Risso's dolphin** (*Grampus griseus*), as well as the **harbor seal** (*Phoca vitulina*) (Au et al., 1997; Kastak et al., 1999). Additional species of toothed whales must be tested because auditory differences among species can be substantial (e.g., porpoises versus killer whales).

There is still no audiogram for any baleen whale species, something that is understandable considering the difficulties of working with these giant species. Since the publication of NRC (1994), several studies have been published that reveal new information on auditory capabilities of seals and toothed whales. Underwater thresholds for three seal species (one eared seal and two true seals) were obtained with lower frequencies than had been tested previously (Kastak and Schusterman, 1998). All three species had relatively high thresholds to sounds with a frequency of 100 Hz (89.9 to 116.3 dB, depending on the species). In addition, Kastak and Schusterman calculated ranges at which the sound from the ATOC source would be just detectable, not necessarily annoying, for all three species: 9 to 34 km for the California sea lion, 160 km for the harbor seal, and 279 km for the elephant seal.[5] Using a different experimental design, Kastak and Schusterman compared low-frequency hearing sensitivity in air and in water for the same three pinniped species. They presented the argument that in water sound pressure is a more biologically appropriate measure than sound intensity.[6] Using sound pressure sensitivity as the critical parameter, they found consistent correlations between hearing sensitivities and the environment in which the animal spends the greater portion of its time. Thus, the elephant seal, which spends a greater proportion of time in water, has better underwater hearing sensitivity than the sea lion, which hears better in air than in water, and the harbor seal, which has almost equal sensitivity in air and in water.

---

[5]Kastak and Schusterman performed these calculations assuming a simplified propagation model of spherical spreading to a distance of 1 km (20 log R), followed by propagation described by 15 log$R$ at greater distances. The authors did not give a reason for using this combination.

[6]Sound intensity is the power of the sound or pressure squared. This raises questions about the biological relevance of comparing in-air and underwater sounds by "correcting for" intensity rather than by simply comparing pressure levels.

Ridgway et al. (1997) tested the hypothesis that the sensitivity of white whale hearing might diminish with depth. To test the effect of depth, two trained white whales made dives to a platform at 5, 100, 200, and 300 m in the Pacific Ocean. During dives to the platform for up to 12 minutes, the whales whistled in response to 500-ms tones projected at random intervals to assess hearing threshold at each of the four depths. Analysis of response whistle spectra, whistle latency to tones, and hearing thresholds showed that the increased hydrostatic pressure at depth changed each whale's whistle response at depth but did not attenuate hearing. Hearing is attenuated in the aerial ear of humans and other land mammals when tested in pressure chambers due to changes in middle-ear impedance that result from increased air density (Fluur and Adolfson, 1966; Pantev and Pantev, 1979; Levendag et al., 1981). The finding that whale hearing is not attenuated at depth suggests that sound is conducted through whale head tissues to the ear without requiring the usual eardrum/ossicular chain amplification of the aerial middle ear. These first-ever hearing tests in the open ocean demonstrate that zones of audibility for human-generated sounds are as great at 300 m, and potentially much deeper, as in shallow water. (These tests could not have been performed without trained whales.)

Au et al. (1997) tested two members of the dolphin family (false killer whale and Risso's dolphin) for their responses to the ATOC signal using a behavioral paradigm. They concluded that neither species could be negatively affected by the ATOC experiments due to their high auditory thresholds (139 and 141 dB, respectively) for the ATOC signal.[7] Based on calculated transmission signal loss, Au et al. concluded that at a horizontal range of 0.5 km these whales would not hear the ATOC signal. They also concluded that the ATOC signal is unlikely to harm baleen whales or to mask their vocalizations, but these conclusions are based on calculations and interpretations of source levels of whale vocalizations without using actual behavioral data. Source levels for baleen whale vocalizations have been calculated from measurements for single animals at considerable distances from the hydrophone. Such vocalization levels may be more appropriately compared to humans shouting, rather than to conversational speech levels. Members of the same species probably are rarely, if ever, exposed to vocalizations at levels as high as the maxima reported by some researchers (cited in Richardson et al., 1995; Au et al., 1997). The reported maxima also may be outliers or erroneous because available data are relatively scarce; thus, assumptions about noise levels that may or may not disturb whales should not be based on such measurements.

---

[7]According to the authors, small toothed whales swimming directly above the ATOC source will not hear the signal unless they dive to 400 m. The authors used a propagation loss model calculated for the source off Kauai. They then stated the horizontal distance without a depth, so it is unclear whether the animals would have to be at the surface for these measurements to be valid.

- **Determine hearing capabilities of larger marine mammals that are not amenable to laboratory study.**

The gray whale has been a favorite species for bioacoustic study because of its predictable occurrence in coastal waters and lagoons. One of the first studies on the response of a baleen whale to industrial noise was conducted on this species (see Richardson et al., 1995 for a review), but no studies on their hearing sensitivity have been conducted. Attempts to obtain auditory evoked potentials from the rescued juvenile gray whale "J.J." while it was recuperating at Sea World (San Diego) had both technical and logistical difficulties because the whale was housed at a facility for public display and no useful data were obtained.

The humpback whale is another species for which considerable acoustic data have been collected, particularly in the Pacific Ocean. A field study was conducted using humpback whale sounds (song, social sounds, and a feeding call) to determine at what levels they respond to other humpbacks (Frankel et al., 1995). A rapid approach response was observed in a small proportion of the playback trials, revealing that the feeding call and social sounds were attractive at distances of 2.8 km. Frankel et al. estimated that the lowest received level that elicited approach was 102 to 106 dB, with a median response level of 113 dB for those sounds. Although limited in scope, this study reinforces the concept that whales may respond to biologically relevant signals near detection thresholds where the signal level is equal to the noise level. These natural sound playbacks evoked responses at lower sound levels than have been reported in the literature for human-generated noise. Interestingly, these values are close to the detection threshold for a white whale vocalization (108-dB broadband measurement) in a captive study using a trained female white whale (Erbe and Farmer, 1998). Thus, the field approach used by Frankel et al. (1995) with baleen whales has provided data that compare well to a more controlled captive study with a toothed whale. The finback whale also shows clear responses to the distant vocalizations of other finbacks in a feeding area (Watkins, 1981), and a similar study on this species could provide additional data to address the issue of relative sensitivity to lower frequencies in a species very likely to be sensitive to low-frequency noise.

Progress in understanding the potential hearing capabilities of whales has been made by theoretical comparisons of the structural components of baleen whale cochleas with those of other marine and terrestrial mammals, particularly elephants (Ketten, 1994, 1997). Ketten (1994) grouped whales into three categories based on cochlear structures: (1) upper-range ultrasonic toothed whales, (2) lower-range ultrasonic toothed whales, and (3) baleen whales.

These analyses can predict the frequency regions at which a species has its most sensitive hearing, but they cannot predict the absolute sensitivity of that ear in the most sensitive frequency region. (Estimates of absolute sensitivity can only be obtained from direct behavioral tests or indirect electrophysiological tests.) If absolute sensitivity can be measured in representative species from each of Ketten's groups, then generalizations could be made to establish "group" data. Because it is impossible to test every species of marine mammal and every type of acoustic parameter with every possible audiometric paradigm, it would be desirable to identify a small but comprehensive set of "model" animals, using the Ketten categorization scheme or the larger scheme proposed in Box 5.1. The hope would be that auditory performance would be similar for species within categories as the acoustic stimuli and auditory tests were varied.

In addition to baleen whales, no tests of hearing capabilities have been published for the sperm whale. Ketten (1994) examined the ear of a sperm whale, but the tissues were not of sufficient quality to conduct the detailed morphological analyses needed to estimate a frequency range. The sperm whale cochlea has characteristics of both the higher and the lower ultrasonic toothed whale cochlea, but the critical measurements of the cochlea and the vestibular system that could indicate whether the ear may be sensitive to the low frequencies of ATOC or LFA sounds have not yet been made.

When investigators are collecting data on hearing capabilities, they should be alert to the possibility of differences in hearing sensitivity as a function of the sex and age of the animal.

Several experiments in which natural sounds were played back to baleen whales appear to have evoked responses that are near the noise background (Malme et al., 1983; Frankel et al., 1995). Behavioral response thresholds in the range of 100 to 110 dB are considerably greater than human underwater threshold hearing levels (Malme et al., 1983; Dahlheim and Ljungblad, 1990; Frankel et al., 1995). As has been suggested by Frankel et al. (1995) and Richardson et al. (1995), field observations of acoustic response thresholds probably have been limited by background noise rather than being indicative of true hearing thresholds. However, these studies have potential for studying sensitivity of hearing in the frequency regions where whales are less sensitive, that is determining the upper and lower limits of hearing this sensitive. These studies can also test the frequencies for which whales have hearing that is noise-limited as opposed to threshold-limited. If particularly quiet sites were found for such studies, it might be possible to obtain lower thresholds. Behavioral methods to study hearing thresholds in large whales using playback of natural sounds should be developed further.

- **Determine audiometric data on multiple animals in order to understand intraspecific variances in hearing capabilities.**

To date, most studies continue to use one or two individuals of a species to determine their audiograms. Some individual variations have been documented in several marine mammals (e.g., harbor seals: Terhune and Turnbull, 1995; Kastak and Schusterman, 1998; bottlenose dolphins: Ridgway and Carder, 1997; white whales: Ridgway et al., 1997). For this reason, measurements from a single animal should be viewed as only a temporary substitute for average hearing capabilities across members of wild populations (such as the bottlenose dolphins that have been studied for several years in Sarasota Bay, Florida). In the absence of large datasets on the hearing capabilities of multiple animals, Terhune and Turnbull suggested a "broad-brush approach" in using available data that assumes average auditory capabilities, for example, rather than using the lowest thresholds measured as standard values. This approach also may be useful when comparing auditory data for various types of whales. Knowledge of the individual differences in hearing sensitivity within species may help explain the large differences observed in the behavior of individual animals (e.g., gray whales) when confronted with a noise source in their path of travel (Tyack and Clark, 1998). Audiograms from many individuals, preferably from the wild, are critically important as a baseline to understanding the hearing capabilities of populations.

- **Determine sound pressure levels that produce temporary and permanent hearing loss in marine mammals.**

In humans, exposure to intense sounds has the potential to produce a number of temporary or permanent aftereffects, depending on the level and duration of the exposure. These aftereffects include reductions in perceived loudness, ringing in the ears (tinnitus), and changes in perceived pitch (Kryter, 1985; Ward, 1997). The most studied aftereffect is loss of hearing sensitivity, commonly known as temporary and permanent threshold shift (TTS and PTS, respectively). Repeated exposures that produce TTS eventually produce a PTS. The fact that some hearing losses last only minutes, hours, or days suggests that some cochlear structures have the ability to recover from whatever damage is inflicted by the exposure stimulus. In terrestrial mammals the receptor cells known as outer hair cells are known to be far more susceptible to acoustic damage than are the less numerous inner hair cells (Saunders et al., 1991), and temporary inactivation of the outer hair cells is presumed to be the primary factor in exposure-induced hearing losses of about 30 dB (in air) or less that last only a few hours. Tremendous variability exists in the susceptibility of individual ears to both TTS and PTS; in some human survey studies the subjects are even partitioned into different groups on the basis of the apparent "toughness" of their ears. For a typical human ear exposed daily to essentially continuous noise in the workplace, 90 dB (in air) has been widely adopted as the point at which precautions need to be taken (see Appendix D). Under U.S. regulations, for every additional 5 dB of exposure, the allowable duration of exposure is halved. However, the 90-dB value (in air) in not ubiquitous across animals and cannot be compared directly with in-water values.

Across mammalian species there are known to be quite large differences in the sound levels required to damage the auditory system. For example, 12 minutes of exposure to a 1,000-Hz tone of 120 dB (in air) sound pressure level can produce more hearing loss in a chinchilla than 12 hours of the same exposure in a squirrel monkey (Hunter-Duvar and Bredberg, 1974). Decory et al. (1992) made direct comparisons of hearing loss and cochlear damage in cat, guinea pig, and chinchilla using several exposure frequencies and a fixed duration of 20 minutes. Across exposure frequencies there was between 10 and 25 dB more hearing loss in the chinchilla than in the cat, with the guinea pig being intermediate between the two. At the highest exposure frequency (8 kHz), there were approximately 10 times the number of damaged hair cells in the chinchilla than in the cat. Because of certain controls implemented by the experimenters, it is possible to rule out any contribution of differences in the outer-ear system to these differences in susceptibility, but it is not yet possible to say with certainty what the relative contributions were from differences in the middle-ear system and differences in cochlear mechanisms.

Decory et al. (1992) presented an interesting argument about the possible contribution to damage made by the angular displacement to which a stereocilium is exposed in each of these species, a suggestion that warrants further study and attention by comparative anatomists studying marine mammals. The work of Luz and Lipscomb (1973) suggests that large discrepancies exist across species for susceptibility to impulse noise as well as continuous noise. There is some evidence to indicate that similar differences in susceptibility to exposure-induced hearing loss exist in marine mammals as well. Kastak and Schusterman (1996) suggested that exposure (in air) to a band level only about 10 to 25 dB above absolute sensitivity was adequate to produce 8 dB of TTS in a harbor seal. By comparison, the work of Schlundt et al. (2000) suggests that for the bottlenose dolphin the sound levels necessary to produce a small masked TTS (in water) must be between about 115 and 150 dB above absolute sensitivity.

The existence of such large differences among species in terms of susceptibility to exposure-induced hearing loss creates two problems for regulators of human-generated sound in the ocean. First, critical exposure levels cannot be extrapolated from a few species, although it may be possible to identify a set of representative species for the initial studies (see Box 5.1). Second, it almost certainly will never be possible to specify one single value of sound level at which damage to the auditory system will begin for all, or even most, marine mammals.

An interesting characteristic of TTS and PTS is that for terrestrial mammals the maximum hearing loss typically occurs in a frequency region above the exposure frequency (McFadden, 1986). Investigators studying TTS in marine mammals should design their experiments to obtain information about any upward shifts in maximum hearing loss that exist in marine mammal ears. In terrestrial mammals, cochlear mechanics are known to be somewhat different in the apical (low-frequency) regions of the cochlea than in the middle and basal regions. Further, some species of bats have cochleas that are highly specialized to process stimuli from certain narrow frequency bands that are important to their survival.

The extent to which these differences affect the growth and spread of TTS and PTS is not completely understood, but because there may be corresponding regional differences in the cochlear mechanics of those marine mammals that depend on low-frequency sounds, it will be important to study TTS following exposures to low-frequency sounds as well as to mid- and high-frequency sounds.

A serendipitous occurrence of TTS was documented in a captive harbor seal exposed to construction noise (Kastak and Schusterman, 1996). Although the exposure was in air, the TTS (also measured in air) nonetheless reveals the levels above threshold at which the seal ear could be affected and the duration of the effect after prolonged repeated exposure. The noise was present over 6 days with 6 to 7 hours of intermittent daily exposure. The longest continuous exposure was believed to be about 1.5 hours. The third octave sound pressure level, centered at 100 Hz, varied from 75 to 90 dB (in air), which was only 10 to 25 dB above threshold for this seal. On day 6 of daily exposures, an 8-dB temporary threshold shift was measured at 100 Hz (the only frequency tested). Also, the seal's false alarm error rate in responding to sounds increased by 23 percent, which the authors suggested may have been the result of exposure-induced tinnitus. Hearing sensitivity recovered completely in about 1 week. Kastak et al. (1999) went on to test underwater TTS in three species of seals. They used test frequencies ranging from 100 to 2,000 Hz and "octave-band noise exposure levels that were approximately 60-75 dB SL (sensation level at center frequency)." They found TTS averaging 4.8 dB for one harbor seal, 4.9 dB for two California sea lions, and 4.6 dB for one northern elephant seal. The animals' hearing returned to baseline levels when tested within 24 hours of noise exposure.

Schlundt et al. (2000) reported measures of TTS in four bottlenose dolphins (three females and one male) following exposures of differing levels at differing frequencies. The exposures were always 1 second in duration because the experiment was designed to determine the effects of a sonar sound commonly used by the U.S. Navy. The behavioral task used was imaginative and requires some description. A platform contained two stations, each having a prescribed place for the animal to position itself in front of a speaker. Under instructions from its trainer, a dolphin would position itself at the first station where it would first be exposed to a sound having a duration of 1 second. Immediately after receiving the first sound, the animal swam to the second station where it was presented with a series of 250-ms tones of fixed frequency and varying level. The animal was trained to whistle whenever it heard a tone from this second speaker, and the level of these test tones was adjusted up and down in accordance with the animal's responses to the series of test tones. After extensive training with this procedure, the level of the 1-second first sound was increased, with the objective of producing 6 dB of hearing loss as detected using the series of second tones. Different animals were tested with the exposure and test tones set to 3, 20, and 75 kHz. In part because of the background level at the test location, the series of second tones was actually masked by a broadband noise whose level could be varied.

Thus, the measure extracted from these data was in fact the difference in the level of the second tone necessary for a fixed level of detection performance in the presence of the background masker. Accordingly, it is probably better characterized as masked temporary threshold shift (MTTS) in order to distinguish it from standard (unmasked) measures of TTS.

This paradigm, while unusual, may provide a more realistic estimate of threshold shifts for animals tested in a noisy environment. The results were that small amounts of MTTS were observed when the exposure sounds were in the range of 192 to 201 dB, and this was true for all three exposure frequencies. (As already noted, this corresponds to exposures approximately 115 to 150 dB above absolute sensitivity, values that are higher than the nominal 90 dB commonly cited for humans.) Further, the animals exhibited varying degrees of behavioral disturbance at exposure levels about 12 to 15 dB below the values necessary to produce MTTS, a fact that Schlundt et al. interpreted to mean that dolphins, and perhaps other whales, may naturally avoid sounds that pose a threat to their hearing. To test this possibility, whales could be monitored closely during the ramp-up phase of experimental presentations of the ATOC signal near the test subjects.

There is a particular priority for obtaining TTS data for endangered baleen whales. Many environmental impact statements (EISs) suggest levels of 150 to 160 dB as safe exposure levels for marine mammals. This is particularly difficult to justify with animals for which no audiometric data exist. Even if there is not enough time to conduct a complete TTS study on a stranded whale, it would still be particularly useful to test for TTS after several tens of minutes of exposure to 160-dB noise.

- **Determine condition of important cochlear structures in wild marine mammals using postmortem examinations** (this topic did not appear in earlier reports).

For many locations around the world the ocean is an extremely noisy place due to shipping, petroleum exploration or drilling, underwater explosions (Ketten et al., 1993), and other human activities. Accordingly, large numbers of marine mammals are already being exposed to anthropogenic sounds on a regular basis, and many of these sounds are of high intensity. Unfortunately, little is known about the auditory consequences of these exposures even though that knowledge could be extremely informative about the possible consequences of exposure to other intense sources such as those used for ATOC and LFA sonar. One obvious source of information would be behavioral or physiological measures of hearing from animals living at these especially noisy sites. In the absence of such difficult-to-obtain information, there is a less direct, but still potentially informative, approach that deserves attention. Specifically, postmortem examinations of the cochleas of marine mammals have the potential to correlate noise levels and hearing damage. Many marine mammals of various species die each year, in strandings, fish nets, and accidents. If the cochleas of some of these animals could be collected, preserved correctly, and sent to scientists capable of histological examination of the cells known to be damaged by exposure to intense sounds, some idea about the current (ambient) level of cochlear damage—and, by implication, hearing loss—gradually would emerge. Eventually, this information could provide a set of baselines for use in establishing new regulations and against which new exposures could be measured.

Much is currently known about the progression of cellular damage that occurs in terrestrial mammals following noise exposures of various types and durations (see Saunders et al., 1991), meaning that the cochlear physiologists have good expectations about what to look for, and where, in the cochleas of marine mammals. Damage of particular types and extents will be definitive evidence of permanent extensive hearing loss, and less severe damage will be evidence of less extensive hearing loss. Of special interest should be the cochleas of young marine mammals because they are the least likely to have cochlear deterioration attributable to factors other than noise exposure (such as the presbyacusis attributable to aging).

In terrestrial mammals, at least, hearing loss induced by a number of agents is known to progress from the basal (high-frequency) part of the cochlea toward the apical (low-frequency) part. Assuming that this pattern of progression exists in marine mammals as well, discovering damage to apical regions of the cochlea that is not accompanied by correspondingly greater damage to basal regions will be strong evidence that the damage was induced by an intense low-frequency sound source. Once frequency maps are established for the cochleas of various representative mammals (Ketten, 1994, 1997), localized damage at a particular region of the cochlea can be traced back to sound sources with maximum energy in a particular frequency band. It is important to undertake a parallel effort to monitor noise exposure in areas where strandings are well monitored. Regular monitoring of noise, strandings, and cochlear damage is required to move research from an anecdotal correlation toward causal links between noise and strandings. When implementing this procedure, the so-called half-octave shift in PTS needs to be taken into account; see McFadden (1986).

Because this idea about gathering information concerning the current state of hearing loss in marine mammals arose late in its deliberations, the Committee was not able to determine how best to implement a program of collecting cochleas from dead marine mammals. Clearly, the SWAT (Standard Whale Auditory Test) teams described elsewhere in this report should include personnel capable of extracting and preserving the peripheral auditory systems of marine mammals. In addition, it might be possible to train other personnel that are likely to come into contact with potential specimens. All interested governmental agencies should be advised about the importance of this collecting work and should be encouraged to implement policies that minimize any existing impediments to its efficient progression. Because of the impossibility of predicting the location of specimens that might become available for this study, and because specimens can deteriorate relatively rapidly, transportation of auditory specimens across international borders for purposes of histological examination should be made as simple as possible. This is especially important for vulnerable endangered species, where the Convention on International Trade in Endangered Species of Wild Fauna and Flora (CITES) permit processes inhibit timely delivery of samples to appropriate experts.

- **Determine morphology and sound conduction paths of the auditory system in various marine mammals.**

No studies have been published since 1994 that reveal new information about the structural details of the auditory system in marine mammals, although ONR has funded a project in the laboratory of Darlene Ketten (Woods Hole Oceanographic Institution) to examine potential auditory pathways in the lower jaw and skull of dolphins as part of a project to model sound conduction in these animals. The existence of a second pathway to the ear may be confirmed by these studies (Ketten, 1994, 1997).

- **Determine temporal-resolving power for various marine mammals** (this topic did not appear in earlier reports).

In humans and other terrestrial mammals, hearing sensitivity varies with signal duration. Specifically, the longer the duration, the less strong the signal needs to be for equal detectability. This relationship holds out to a limiting duration of about 300 to 500 ms, beyond which the signal level for equal detectability remains constant. What is most interesting about this trade-off between duration and level is that level declines (or increases) about 3 dB for every doubling (or halving) of duration below the limiting duration. That is, power is traded almost perfectly for time (e.g., Plomp and Bouman, 1959), suggesting that energy is the feature of the signal that determines detectability. This temporal integration function rises about 15 dB as the signal duration decreases from 300 ms to 10 ms in any animal using signal energy as the basis for its performance.

The implications of temporal-resolving power differ for marine mammals with low-frequency hearing versus those with high-frequency hearing. Malme et al. (1983, 1984) reported that the levels required to elicit a response in gray whales need to be as much as 50 dB greater for short impulses than for continuous sounds. These data suggest that this species may be disproportionately insensitive to very short sounds; that is, their temporal integration function appears to be much steeper or displaced from that of typical terrestrial mammals. Data on the hearing capabilities of other baleen whales may suggest that long temporal integration functions, with a corresponding relative insensitivity to transient sounds, is to be expected in other baleen whales. One implication is that these species should have generally poor temporal resolution; for example, they should have difficulty detecting a brief silent gap in an ongoing waveform.

In contrast to baleen whales, characterized by their sensitivity to low-frequency sounds, the toothed whales, characterized by high-frequency hearing, do appear to exhibit short temporal integration functions. Measures of temporal resolution capability and temporal integration times were obtained in several species of toothed whales by Dolphin (1995, 1996) using evoked potential techniques. The species examined demonstrated temporal resolution capabilities exceeding those observed in humans and other terrestrial mammal species. Thus, one would expect these species to exhibit high sensitivity to brief, transient sounds. Moreover, the possession by dolphins of the capability for both very high temporal resolution and extremely sharp frequency resolution is enigmatic and points out our lack of understanding of the auditory processing of these animals. Accordingly, it would be extremely valuable to obtain information about temporal processing in a wider range of species from each of the groups identified in Box 5.1.

Although the evidence suggesting differences in temporal processing capabilities between certain marine mammals and terrestrial mammals is unquestionably scarce, a logical argument can be made for such a difference. Species that are highly dependent on low frequencies may have acquired long time constants for temporal integration, making them highly insensitive to very short sounds (the rise time of their auditory systems may be so long that the response to impulsive sounds is weak), whereas those species that depend on high frequencies and/or echolocation have evolved auditory systems with short time constants and high temporal resolution capabilities. If this is true, it would have major impact on what kinds of sounds are regarded as dangerous for different species; namely, it may be that brief sounds, even quite intense ones, are not unduly dangerous to hearing in some species of baleen whale. These facts emphasize that temporal resolution can vary substantially across species having hearing that is specialized for operation in different frequency regions and they lend plausibility to the speculation above about temporal resolution possibly being poor in those marine mammals specialized to communicate with low-frequency sound.

EFFECTS OF LOW-FREQUENCY SOUNDS ON THE FOOD CHAIN

Aim: To determine whether low-frequency sounds affect the behavior and physiology of organisms that serve as part of the food chain for marine mammals (NRC, 1994, pp. 53-54).

The most serious effects of noise on potential prey species are those that involve growth and reproduction. Increases in noise (above ambient levels) have been implicated in reduced growth and reproduction in a variety of marine organisms in tanks, although these data are still very limited and in need of replication. A single study of the effects of intense sound on the auditory systems of freshwater fish has been published (Hastings et al., 1996). In this study, freshwater fish were exposed to varying levels of a sinusoidal sound, and some evidence of auditory damage was reported for sounds above 180 dB when fish were exposed to continuous signals for 4 hours and then allowed to survive for several days before damage was assessed. It should be noted that these sounds were very different in intensity, duty cycle, and duration from those used in the MMRP studies and the fish used in this experiment could not escape the sound source. Additional studies are needed, particularly of fish species that are endangered, important commercially, or are a component of the food chain of marine mammal species.

Growth rates in two fish species, **sheepshead minnows** (*Cyprinodon variegatus*) and **killifish** (*Fundulus similis*), were significantly lower in aquariums exposed to noise 20 to 30 dB above ambient levels in the natural habitat (Banner and Hyatt, 1973). Tanks with noise levels 20 dB higher than ambient levels reduced the viability of eggs in sheepshead minnows. Evidence reviewed by Corwin and Oberholtzer (1997) suggests that fish and perhaps some shark and amphibian species have the capacity to regenerate damaged hair cells in their auditory and balance organs (see also Lombarte et al., 1993). To the extent that this ability is widespread in those species, they may be at lesser long-term risk from exposure to intense sounds than are marine mammals, which are presumably like terrestrial mammals in being incapable of regenerating new receptor cells to replace damaged ones (e.g., Hastings et al., 1996; Corwin and Oberholtzer, 1997).

No assessments of pre-ATOC shark abundances were made, nor has the potential attraction of sharks by low-frequency ATOC sound been investigated, despite extensive data in the literature showing that low-frequency sounds, such as those used by ATOC, attract sharks (e.g., Myrberg, 1972, 1978; Myrberg et al., 1976). The potential for redistribution of sharks cannot be ignored, and some effort should be made in the future to monitor any ATOC source with appropriate methods (methods that would not alter the behavior of sharks and other organisms in significant ways) to determine if sharks are attracted to the site.

Effects of intense sound have been observed in a shrimp species (*Crangon crangon*) (Lagardére, 1982). Shrimp exposed to noise levels 20 to 30 dB higher than normal ambient levels exhibited reduced growth and reproduction and increased aggression and mortality relative to a control group.

Although population surveys have indicated the presence of endangered turtle species at both ATOC sites (the Kauai EIS mentions green sea turtles [*Chelonia mydas*], leatherback turtles [*Dermochelys coriacea*], olive Ridley turtles [*Lepidochelys olivacea*], and hawksbill turtles [*Eretmochelys imbricata*]) no specific results for studies of hearing or behavioral observations on any shark or turtle species were presented by the MMRP scientists.

POTENTIAL NONAUDITORY ACOUSTIC EFFECTS ON MARINE ANIMAL HEALTH

Extremely intense sound can result in injury to many bodily organs and physiological processes. Aside from the direct effects of nearby blasting from explosives, intense sound can cause injury to lungs and other air-containing spaces. In addition, there may be direct effects on the nervous system. Extremely intense sound can also affect the vestibular system and can cause disorientation. In humans these vestibular impacts result in readily observable *nystagmus*, an uncontrolled movement of the eyes (Stephens and Ballam, 1974). Studies of these effects in humans and other terrestrial species are being funded by ONR.

Research by Crum and Mao (1996) suggests the possibility of potential effects that only occur in certain physiological states that would be very difficult to study in nonhumans. For example, human divers are susceptible to decompression sickness ("bends"), a disabling and sometimes fatal condition in which bubbles of nitrogen gas form in the blood, joints, and other tissues. Low-frequency sound might induce bends episodes in human divers whose tissues are saturated with gas due to breathing pressurized gas at depth, that would not otherwise occur. Crum and Mao (1996) showed experimental data suggesting that intense (160 to 220 dB) low-frequency sound may induce bubble growth in tissues (see also Lettvin et al., 1982), and therefore divers ensonified with low-frequency pulsed sound when they are near decompression limits could be severely injured (Crum and Mao, 1996). This is unlikely to be a problem with the ATOC sources because they are so far offshore from dive sites and in deep water, but it may be a problem with more powerful shallow-water sources, such as SURTASS-LFA.

Although marine mammals do not carry a tank of pressurized breathing gas as human divers do, they make frequent dives to depths greater than 100 m, which may produce over 200 percent supersaturation of nitrogen in muscle tissue after repetitive dives (Ridgway and Howard, 1979, 1982). In Ridgway and Howard's 1979 study, dolphins made 23 to 25 dives to 100 m (10 atmospheres) in 1 hour. Dolphins did not suffer from decompression sickness even with muscle nitrogen at supersaturated levels that would produce bends in humans (Ridgway and Howard, 1979). However, Lettvin et al. (1982) and Crum and Mao (1996) suggested that sound exposure could induce bubble growth in blood. Crum and Mao (1996) suggested that this might be an issue for both humans and marine mammals. Therefore, it should be considered whether intense low-frequency sound might cause bubbles in the circulatory systems of whales returning to the surface after a long series of deep, but rapid, dives similar to those studied by Ridgway and Howard (1979). At the present time, bubble formation in diving marine mammals must be considered as conjecture based on findings in unrelated studies. However, research on marine mammals should probably consider the issue after ONR studies on human subjects have been completed. If results from research on humans suggest that marine mammals could experience such effects, sophisticated electronic tags that record depth of dive and other relevant data might reveal the extent to which such a nonauditory threat may be realistic in wild populations.

As part of the SURTASS-LFA program, ONR funded research on the effects of low-frequency sound on human divers. These studies show that the likelihood of scuba divers aborting a dive is lowest at 250 Hz but rises for source frequencies both above and below 250 Hz. Sound levels were relatively nonaversive until reaching a received sound pressure level of 148 dB, at which point 15 percent of dives were aborted. Aversion at lower frequencies resulted from a sensation of vibration in the air-filled cavities in the head, chest, and abdomen. No vestibular effects were observed for sound up to 157 dB.

Research on marine mammals should also be devoted to evaluating other physiological measures of general health, both short-term and long-term, that can be monitored non-invasively over time with tags. For now, these measurements are limited to heart rate and respiration. Data on stress and stress indicators in marine mammals is sorely lacking, and there is not even a baseline from which to determine normal values. At the same time, the significance of such reactions to stress in terms of reproduction and survival should be assessed.

LONG-TERM ACOUSTIC MONITORING OF CRITICAL HABITATS

The issue of cumulative impacts from human-generated noise is best dealt with as a habitat degradation issue. Undersea noise should be monitored in important marine mammal habitats (after these have been identified). This monitoring effort should be designed in parallel with surveys of marine mammal distribution, abundance, and strandings using methods that allow temporal and spatial analysis of how noise may lead to changes in these population characteristics. These data are particularly important for populations that are either not recovering or are declining due to habitat degradation and other causes. Monitoring should include the ambient noise field, marine mammal vocalizations, and transient noises, particularly in locations and times of the year in which marine mammals are known to be common. This monitoring optimally should also include or be coordinated with other assessments of habitat quality such as prey fields and chemical pollutants. Coordination of noise monitoring with stranding networks would enable more systematic and controlled evaluation of whether noise influences strandings and whether cochlear damage in stranded animals is associated with acute noise exposure.

NMFS, the Navy, and other agencies with responsibilities for marine mammals or that conduct or permit activities that introduce significant levels of sound to the ocean should evaluate the costs and benefits of an array of acoustic receivers designed to monitor both human-generated sound in the ocean and the vocalizations of whales in acoustic hotspots (NRDC, 1999). One possibility is to use existing arrays such as the IUSS (JOI, 1994; Clark, 1995; Gisiner, 1998) developed by the U.S. Navy to detect submarines. Evaluation of the appropriateness of the IUSS should determine whether the bandwidth and geographic coverage of the IUSS is adequate for the task of monitoring ambient noise and marine mammals or whether it is necessary to design and build an array of sensors specifically to monitor marine mammals. Such a system could be automated to activate when significant sounds are detected. Whales could be located and tracked in real time and in three-dimensional space, thus identifying natural paths and avoidance paths. This capability was demonstrated in the Whales '93 program in which the IUSS was used to routinely detect, locate, and track blue, finback, and humpback whales in the North Atlantic Ocean (JOI, 1994). Hundreds of thousands of whale vocalizations were documented, allowing the description of seasonal movements of the whales. Autonomous underwater recorders, sonobuoys, or towed arrays of hydrophones can be used in areas where (or at times when) more intensive monitoring is desired (Richardson et al., 1986; Thomas et al., 1986; Moore et al., 1999).

4   Regulatory Issues

ACOUSTIC HARASSMENT

The intent of the Marine Mammal Protection Act (MMPA) is the management and regulation of human activities that affect distinct populations of marine mammals and the protection of essential marine mammal habitats. Conservation of whales and seals and their environments must be continuous, and the National Marine Fisheries Service (NMFS)[1] should develop long-term strategies to fulfill the MMPA's mandates. For example, NMFS has an obligation to collect the necessary data to monitor the long-term health of whale and seal populations, including population trends, distribution and abundance, and definition and status of critical habitats.[2] Conserving populations and habitats should be the guiding principles for regulation of activities impacting whales and seals. The acoustic parameters of the habitat are just as important as other habitat characteristics, although much less is known about the acoustic features of critical habitats. The National Research Council (NRC, 1994) devoted one of its three chapters to marine mammal regulatory issues. This chapter discusses the existing regulatory procedures affecting acoustic harassment, recommends changes to those proce-

---

[1]The U.S. Fish and Wildlife Service is responsible for conservation for some marine mammal species, including manatees, dugongs, polar bears, walruses, and marine and sea otters. The Committee does not consider such species in this report because most are less likely to be affected specifically by low-frequency sound than are whales and seals, although the vocalization and hearing capabilities of these species have not been well characterized. NMFS has the responsibility for whales and seals (except walruses).

[2]The Endangered Species Act requires the designation of critical habitats for endangered and threatened species.

dures, and discusses the permitting process. Here, the Committee considers how the MMPA could be changed in its next reauthorization to improve the definition of harassment from acoustic sources.

The NRC (1994) suggested that the regulations governing the taking of marine mammals by fishing activities should be broadened to include other user groups that might take marine mammals. This concept was incorporated into the 1994 MMPA amendments. The MMPA now requires calculation for each species of a conservative number of animals that might be taken by humans from marine mammal stocks, while "allowing that stock to reach or maintain optimum sustainable population," called the *potential biological removal* (PBR) level (MMPA, Sec. 1362[t]; see Appendix C). NMFS is required to tally all human-induced mortality for its stock assessments (MMPA Sec. 1386[a]) and uses this number to estimate PBR. The PBR regime seems to be working, although additional effort may be required to quantify marine mammal takes from all human activities so that they can be incorporated into the PBR, even though this might reduce the takes allowed to fishermen. However, it would be virtually impossible in the near future to sum all sources of lethal takes to compare with the PBR for any species and lethal takes by sound would be particularly problematic because it is so difficult to prove that they resulted as a consequence of human-generated sound. Takes are not allocated officially, but many fisheries operate under prohibited species quotas and thus automatically received part or all of the PBR level of takes, whereas research scientists must apply for takes through a small-take exemption. If the takes counted against the PBR of a marine mammal stock are known to not include all human takes, the PBR should be adjusted downward accordingly.

The core of the MMPA is a "moratorium on the taking or importation of marine mammals" (Sec. 1371). "Taking" is defined in the MMPA as "to harass, hunt, capture, or kill or attempt to harass, hunt, capture, or kill any marine mammal" (Sec. 1362[m]). The 1994 amendments to the MMPA included a definition of harassment (Sec. 1362[r]) as "any act of pursuit, torment, or annoyance which:

Level A—has the potential to injure a marine mammal or marine mammal stock in the wild; or

Level B—has the potential to disturb a marine mammal or marine mammal stock in the wild by causing disruption of behavioral patterns, including, but not limited to, migration, breathing, nursing, breeding, feeding, or sheltering."

In its 1994 report the NRC identified drawbacks to these definitions. Swartz and Hofman (1991) reviewed the issue of taking by harassment in the context of small-take authorizations before the enactment of the 1994 MMPA amendments. They noted (pp. 2-3) that "the term 'harass' has been interpreted through practice to include any action that results in an observable change in the behavior of a marine mammal—for example, abrupt termination of breeding or feeding, avoidance behavior, and changes in swimming speed, dive frequency, dive duration, or direction of movement." The NRC noted that, as techniques for observing marine mammals improve, it may become possible to observe responses as soon as an animal can detect an acoustic signal, even though such responses may not constitute evidence of a significant negative effect. This has, in fact, occurred. For example, Burgess et al. (1998) were able to track the heart rates of free-ranging elephant seals, and time-depth recorders recorded subtle meter-by-meter patterns of dive behavior. Such data can be combined with data on received sound levels to determine behavioral thresholds of sound, levels that animals react to with some physiological or behavioral response, but which are not necessarily dangerous to the animals. Conversely, for long sound exposures (hours to days), TTS can occur without any behavioral response. It cannot be assumed that avoidance responses to continuous noise will prevent injury in the wild. For example, animals might be motivated to approach a loud source that produced a TTS if the source was near food that the animal sought. This reinforces the need to focus on predicted TTS rather than behavior, although this is not now possible for most species. The NRC (1994, 1996) has advocated a regulatory definition of harassment that focuses on adverse effects to marine mammals.

The Committee supports this effort to distinguish between injury and disruption of behavior and proposes a refinement of the above definitions to incorporate and differentiate between immediate injury and longer-term, significant physiological and behavioral effects that may affect the growth, reproduction, or mortality of animals. Moreover, regulatory efforts directed at minimizing and mitigating the effects of anthropogenic sounds on marine mammals and other marine organisms should have the goal of minimizing the risk of injury and meaningful disruption of biologically significant activities, where biological significance is defined as having potential demographic effects on reproduction or longevity.

Definition of Level A Acoustic Harassment

The definition of Level A acoustic harassment should be related to the likelihood that a sound will produce a temporary threshold shift (TTS), as well as to the magnitude of the TTS. However, relatively little is known about TTS in marine mammals, and this would be a difficult standard to implement, at least with existing knowledge. The problem of using TTS as an absolute standard of injury is illustrated by terrestrial mammals, for which it is possible for an animal to exhibit small amounts of TTS on numerous occasions without TTS developing into a PTS. Animals that experience only low levels of TTS are not going to be injured, suggesting TTS as a conservative standard for prevention of injury. In humans the best predictions about the relationship between TTS and PTS come from datasets involving noise exposures in the workplace. For those situations the standard rule of thumb is that the amount of TTS exhibited at the end of a single workday will become a PTS of that same magnitude after approximately 10 years in the workplace (e.g., Nixon and Glorig, 1961). However, that prediction is based on the assumption of daily exposures to that same sound for 5 days per week and 50 weeks per year during that 10-year period. That is, this prediction presumes regular long-duration exposures. With less frequent exposures the time required to cause equivalent PTS would be extended. Clearly relevant here for the question of Level A harassment is the amount of TTS produced because large amounts of TTS will lead to measurable amounts of PTS sooner than will small amounts. That is, there might be a negative effect on the ability of a marine mammal to hear and communicate after only a few exposures to sounds strong enough to produce large amounts of TTS in that species. (As noted elsewhere in this report, maximal TTS and PTS often occur at frequencies above the exposure frequency by as much as an octave or more.)

For certain animal models it appears that TTS of 10 dB and less within 15 minutes after the exposure is fully reversible and without obvious cochlear damage (Liberman and Dodds, 1987; Ahroon et al., 1996) as long as the exposures are not continued for long periods of time. In both these studies, cochlear damage was evident only after TTS exceeded 40 to 60 dB within 15 minutes after the exposure. However, species differences are relevant here, too, because there is some evidence in primates that even in the absence of a measurable PTS or a functional change in hearing, there can be anatomical evidence of damage to the cochlea following repeated episodes of TTS over the course of months (Lonsbury-Martin et al., 1987).

These facts, coupled with the general ignorance that exists about TTS in marine mammals, make it impossible to identify an exposure level that would be unequivocally safe for all the members of a species. However, as a preliminary criterion, it seems reasonable to presume that any sound that produces a TTS of 10 dB or less in exposure episodes that are separated by nonexposure intervals that are ample to allow full recovery (at least 24 hours) does not constitute a major risk to the auditory system of a marine mammal. As knowledge of the auditory systems of marine mammals increases, this preliminary criterion should be reexamined.

Definition of Level B Acoustic Harassment

It does not make sense to regulate minor changes in behavior having no adverse impact; rather, regulations must focus on significant disruption of behaviors critical to survival and reproduction, which is the clear intent of the definition of harassment in the MMPA. For example, Malme et al. (1983) documented that migrating gray whales show a statistically significant avoidance of an area a few hundred meters around a source playing back the sounds of oil industry activities. It is difficult to assess the impact of this avoidance on gray whale survival and reproduction because the adaptive value of migrating close to shore is unknown.

From one perspective, this avoidance is an adaptive response because it would protect the whales from approaching industrial activities too closely and would, at worst, delay their migration from Alaska to Mexico by a few minutes. From another perspective, even a small avoidance could result in unanticipated consequences. If the avoidance response makes the whales more vulnerable to predation by killer whales, or the sound producing the avoidance response masks killer whale sounds or other environmental cues, the avoidance could have effects beyond delaying migration.

Activities that produce statistically significant but biologically insignificant responses are subject to take authorizations under the MMPA and ESA as the regulations are currently implemented; responsible agencies must provide authorization unless there is good justification for concluding that the effects will not be negligible. Such review would be a reasonable approach if "negligible effects" were defined more appropriately. For example, current research suggests that thousands of ships each day are likely to cause short-term avoidance responses, and many of these responses may help reduce the risk of vessel collision. If the current interpretation of the law for Level B harassment (detectable changes in behavior) were applied to shipping as strenuously as it is applied to scientific and naval activities, the result would be crippling regulation of nearly every motorized vessel operating in U.S. waters. NMFS should promulgate uniform regulations based on their potential for a biologically significant impact on marine mammals. Thus, Level B harassment should be redefined as follows:

> Level B—has the potential to disturb a marine mammal or marine mammal stock in the wild by causing meaningful disruption of biologically significant activities, including but not limited to, migration, breeding, care of young, predator avoidance or defense, and feeding.

The Committee suggests limiting the definition to functional categories of activity likely to influence survival or reproduction. Thus, the term "sheltering" that is included in the existing definition is both too vague and unmeasurable to be considered with these other functional categories.

There are several exemptions to the moratorium on taking marine mammals provided in the MMPA and its implementing regulations. They include (1) permits for scientific research on marine mammals (Sec. 1374[c][3]); (2) authorization by rule making to take small numbers of marine mammals incidental to activities other than commercial fishing (Sec. 1371[a][5][A]); and (3) authorization by a more streamlined process for the unintentional take of small numbers of marine mammals by harassment incidental to activities other than marine fishing (Sec. 1371 (a)(5)(D)). Each exemption has implications for uses of sound for scientific research in the ocean. Although Congress intended to provide less stringent means for marine scientists to obtain permission to unintentionally harass marine mammals to an insignificant degree, NMFS has applied its regulations most stringently to science.

The criterion of negligible impact on a population from all human activities should consider the number of individuals (or percent of population) potentially impacted and the risk of impact to each individual. Decisions should consider critical habitat issues and the status of a population, as well as sensitivity of different marine mammals to the type of activity proposed. For activities that include introduction of sound to the ocean, frequency, duration, temporal characteristics of the sound, and the relevance of these sound qualities to characteristic species should be considered. Also, the total duration and spatial extent of the sound field must be taken into consideration (Reeves et al., 1996). The ultimate long-term goal should be a risk function involving intensity and duration of exposure (see Miller, 1974) for each species, but our current lack of knowledge impedes this goal.

Scientific Research Permits

When Congress reauthorized the MMPA in 1994 it allowed the issuance of general authorizations for research on marine mammals. NMFS has excluded acoustic studies not focused on marine mammals (like the Acoustic Thermometry of Ocean Climate experiment) from this harassment authorization category. The rationale for this exclusion is that permits are for research *on* marine mammals; research that incidentally *affects* marine mammals was not meant to be covered.

The Committee believes that all forms of scientific research permits, relating to both Level A and Level B harassment, should be judged by compatible standards. The existing regulatory regime does not consistently regulate research that could affect—either directly or indirectly—marine mammals. For example, a biologist proposing to study how a whale responds to vessel noise would have to apply for a scientific research permit, whereas an oceanographer planning to transit the same habitat in a large research vessel would not be subject to any regulation, and an acoustician using a similar level of sound for studies unrelated to marine mammals might need to obtain an incidental harassment authorization. It seems illogical to regulate the artificially induced acoustic stimuli more intensely than the vessel-induced sound, which adds the risk of actually striking the whale. For example, the MMPA and NMFS regulations should include acoustic studies in the regulatory procedures related to approvals for harassment during scientific research.

Incidental Harassment Authorization

NMFS has proposed regulations for implementing Section 1371(a)(5)(D) of the MMPA, which provides a streamlined process for obtaining incidental take authorization when the taking would be by harassment only (DOC, 1995). These regulations appear to address some of the concerns of NRC (1994), which stated that regulation of acoustic harassment posed significant barriers to scientific research. An interim final rule was published in 1996 to expedite the processing of incidental harassment authorization (IHA) requests for oil- and gas-related activities (DOC, 1996). NMFS expects to publish new criteria in the near future and will develop new regulations and guidelines after review and comment on the criteria. The delay in issuing final regulations occurred because NMFS, the Navy, and other groups wanted to discuss the effects of noise on marine mammals in workshop settings to gather a broad base of information (e.g., Reeves et al., 1996; Gisiner, 1998).

Persons requesting an IHA must provide information demonstrating that any taking is likely to be by harassment only, will be unintentional, will involve small numbers of marine mammals, and will have a negligible impact on the affected species and stocks. NMFS has defined "small numbers" as a portion of a marine mammal species or population stock whose taking would have negligible impact on the viability of that species or stock. The Committee supports incorporating population status into regulations on harassment. The duration and expected severity of the proposed harassment also should be factored into these authorizations, recognizing that the environmental impacts of acoustic pollution, like other environmental impacts, can have cumulative effects.

In addition to making the suggested change in the Level B harassment definition, it would be desirable to remove the phrase "of small number" from MMPA Section 1371(a)(5)(D)(i). If such a change is not made, it is conceivable under the current MMPA language there would be two tests for determining takes by harassment, small numbers first, and if that test were met, negligible impact from that take of small numbers. The suggested change would prevent the denial of research permits that might insignificantly harass large numbers of animals and would leave the "negligible impact" test intact.

CUMULATIVE IMPACTS

Even if marine mammals are protected on a case-by-case basis from individual acts of harassment extreme enough to have an adverse impact, they may require additional protection from milder harassment that is repeated so often that impact accumulates. One way to address this issue is to study how animals respond to repeated exposure. Most animals habituate[3] to repeated exposure to the same stimulus. This reduced probability and reduced intensity of response suggest that applying a response model based on single exposures may be overly conservative, but there are few data on habituation in marine mammals.

Alternatively, animals can be sensitized to stimuli, in that their later responses may be greater than earlier responses. The Committee is not aware of any

---

[3]Behavioral habituation is "a desensitization to a specific stimulus situation" (Lorenz, 1981). Habituation to a loud noise is not necessarily adaptive and could actually make TTS or PTS more likely.

definitive experiments conducted since 1994 on habituation of marine mammals to a noise stimulus, although the National Oceanic and Atmospheric Administration and the Marine Mammal Commission are extremely interested in this issue because of its relevance to using higher-frequency sound (e.g., 10 kHz) as an acoustic deterrence to reduce interactions between marine mammals and the fishing industry (Reeves et al., 1996). Kastak and Schusterman (1996) showed in unpublished research that one individual harbor seal and two California sea lions eventually habituated to clicks and frequency-modulated tone stimuli, but an elephant seal appeared to become sensitized. Clearly, these data are limited by our ignorance of differences among these species, but the Kastak and Schusterman study does underscore the importance of investigating species differences in sensitivity to acoustic stimuli (see Chapter 5) because there may be species-specific responses to representative anthropogenic sounds.

The Committee also suggests that activities that are presently unregulated, but which are major sources of sound to the ocean (e.g., commercial shipping) be brought into the regulatory framework of the MMPA. Such a change should increase protection of marine mammals by providing a comprehensive regulatory regime for acoustic impacts on marine mammals, eliminating what amounts to an exemption on regulation of commercial sound producers and the current and historic focus on marine mammal science, oceanography, and Navy activities.

Findings and Recommendations

The Committee discussed a wide range of topics related to its charge. This chapter presents the results of the Committee's review of the Marine Mammal Research Program (MMRP), identification of important future research and observations, specification of desirable regulatory reforms, and ideas for comprehensive monitoring and regulation of sound in the ocean.

RESULTS OF THE MMRP

Findings: Tests of the Acoustic Thermometry of Ocean Climate (ATOC) source were authorized under permits for the MMRP experiment. Although the MMRP was allowed some control over the specific days that the California transmissions took place, MMRP was retrofitted to an operational use of the ATOC source and was not designed to investigate the effects of the source most effectively. This situation illustrates problems that can be encountered in studies designed to monitor animals in the area where a noise source has been placed and is operated for other reasons, rather than optimizing transmissions to accomplish a specific research objective. As a consequence, the results of the MMRP do not conclusively demonstrate that the ATOC signal *either* has an effect *or* has no effect on marine mammals in the short- or long-term. In view of the lack of data for marine mammals exposed to the ATOC signal at received levels above 137 dB, and the incomplete analyses of much of the data collected off the Kauai source, the Committee could reach no conclusions about the effects of the ATOC source at the level of 195 dB. Data from tests of oil industry noises (Malme et al., 1983) and the low-frequency active (LFA) sonar tests (Tyack and Clark, 1998) indicate that these kinds of signals can alter the normal behavior of migrating gray whales, and data from MMRP are suggestive of a behavioral change in humpback whales and sperm whales exposed to 130 dB or greater from the ATOC signal. Thus, there is a potential for altering the normal behavior of marine mammals capable of hearing low-frequency sounds, such as those produced by the ATOC source, LFA sonar, and commercial shipping. The biological significance of short- and long-term exposure cannot be extrapolated from the limited data acquired during the short-term MMRP studies. Redistribution of marine mammals from traditional feeding or breeding areas was not observed, but cannot be ruled out.

Recommendations: If ATOC does continue, a marine mammal monitoring and research component should be required, but the marine mammal research program should have the flexibility to design and perform playback experiments optimized to yield information about biologically significant effects of the source on marine mammals. In general, any long-term experiment involving high source level ensonification of large areas of the ocean should take precautions to minimize exposure of marine mammals to dangerous levels of sound. Such precautions could include one or more of the following:

• Design the physical structure of the source to minimize the potential exposure of marine organisms, if this is technically feasible. For example, a physical barrier could be erected around the source, like a radome on a radar facility.
• Install sensors to shut down the source automatically when marine mammals are too close.
• Make the source level and duty cycle as low as possible, given the objectives of the research.
• Install the source away from large concentrations of marine mammals.
• Design the ramp-up period to make it possible for marine mammals to avoid received levels that would cause temporary threshold shift (TTS).

If the Kauai ATOC transmissions are continued, it will be important to continue ship- and air-based measurements of marine mammals within the 130-dB zone around the source. The Committee was told that ATOC investigators plan to continue aerial observations. Observations of marine mammals near the Kauai source should include humpback whales as well as other species. Vessel-based observations and aerial surveys are complementary and both methods should be used. The inability of shore-based observers on Kauai to observe humpback whales near enough to the ATOC source to be exposed to sound levels that would be likely to produce biologically significant behavioral responses indicates that shore-based observations should not be continued for studying the effects of the Kauai source. Shore-based methods are useful, however, when animals exposed to the sound levels of interest can be viewed easily from shore (e.g., Malme et al., 1983, 1984; Frankel and Clark, 1998a).

Long-term observations should be conducted near the Kauai source. Priority should be given to specific studies targeted to resolve areas of critical uncertainty. Since the source may be deployed for at least 5 more years to meet the ATOC objectives, long-term studies of animals near the source are a priority. If the source is to be operated for decades as part of an operational ocean-observing system, it is important to determine whether there are some animals that are resident in the area, because in such a case small effects might accumulate over time to have a larger impact. A study should be designed specifically to determine whether there are resident animals in the source area and to study whether the ATOC source changes the pattern of residency for identified individual animals on day-to-day and interannual timescales. Such observations would be more sensitive indicators of habitat usage for resident populations than the more general comparisons of the distributions of sightings of each species conducted by the MMRP.

1. A vessel-based study to photo-identify marine mammals sighted near the source could be used to test whether there are any resident individuals or populations. If carried out over the years ATOC hopes to operate, such an observation program could provide more information on the status of the population (as did similar data for baleen whales off California; Calambokidis, 1996, 1999). The vessel conducting this photo-identification work could also combine visual and acoustic monitoring of marine mammals and monitor for any unexpected effects of ATOC transmissions with a statistically designed sampling program. In addition, ATOC has a responsibility to design a well-controlled study on the effects of ATOC transmissions on humpback whales within the 130-dB zone around the source. This would probably need to be vessel-based in order to follow animals near the source. It could use the same vessel involved with the photo-identification work.

2. Conduct aerial surveys around the ATOC source. If aerial surveys are conducted early enough in the season to document the migration of humpback whales into the area, there will be some evidence as to whether individual animals enter the area around the receiver and later are repelled by the source. Individual animals are impossible to track for long from an airplane, but a broader coverage can usually be accomplished using aircraft rather than shore- or ship-based methods. Aerial surveys also allow documentation of the distributions of sperm whales and other species that are more difficult to observe from vessels to determine whether there is cause for concern about long-term exposure of these species.

3. Use radio-tagging and tracking, and recoverable data recorders. The use of such tags to study the effect of the California source on elephant seals provided a wealth of data on diving behavior and received levels of sound. Although tags and data recorders are harder to place on whales than on elephant seals and harder to retrieve later, they could be useful in providing the same kind of information

on received level and diving behavior, with a potential for recording whale vocalizations.

4. Conduct studies on the effects of ramped-up signals. Ramped-up signals are used worldwide in high-energy seismic surveys as a common-sense measure, but their effectiveness has not been tested scientifically. Funding should be designated from the ATOC program or other sources for competitive grants to study the responses of seals and whales to a ramp-up of a noxious sound. Since baleen whales and sperm whales cannot be tested in controlled studies in captivity with the ATOC signal, and since most other species probably cannot hear it, a higher-frequency ATOC-like sound could be used. For example, responses of captive dolphins or white whales could be used to document the kinds of responses or lack thereof to a ramped-up signal. Do they ignore it or do they approach to investigate the new sound? How many exposures does it take (if ever) for them to leave the area as the ramp-up begins? Do they habituate to the ramped-up signal? At what level should the ramp-up start, in order to avoid startling the animals but still be heard above background noise (e.g., should the ramp-up phase start with a lower source level than 165 dB)? This would not be a perfect experiment because it could be argued that baleen whales might not respond the same way as small toothed whales, nor wild animals like captives, but such an experiment should provide insight into the potential behavior of other marine mammals. The Minerals Management Service (MMS) is funding a study in the Gulf of Mexico region of signal ramp-up that is designed to repel small toothed whales and seals before seismic surveys using airguns are started. Data from this study should be reviewed by ATOC investigators.

FUTURE RESEARCH AND OBSERVATIONS

The Committee reiterates the research recommendations made in NRC (1994, 1996). Although progress has been made on many of these topics (e.g., TTS), all remain valid and were discussed in detail in Chapter 3. In addition to these research topics, research priorities are identified below and suggestions made for the appropriate institutional structure for selecting, funding, and overseeing research. The federal agencies and Congress should determine whether these activities are of high enough priority to merit reprogramming of existing funds or appropriating new funds. The speed with which these research and observation activities are implemented will depend on the urgency felt by the public, Congress, and federal agencies for gaining the knowledge necessary to address the goals of both protecting marine mammals and carrying out desirable human activities that add sound to the ocean.

Priority Studies

Findings: The typical pattern for funding research on the effects of noise on marine mammals is for a group responsible for producing noise to be required to provide data on the effects of its operations. This leads the group to fund projects closely tied to the specific signals, areas, times of planned operations, and species of special concern to that project. For example, the primary approach of the MMRP has been to study one anticipated sound stimulus at three anticipated source sites. A problem with this approach is that results cannot be extrapolated readily to other stimuli or sometimes to other sites. There are hundreds of sources of potential concern, and it may be more efficient to develop a more comprehensive research program using a set of representative stimuli to more easily allow any users to model the predicted response to their own stimulus.

Sound is multidimensional and cannot be characterized fully by a single measure, for example, peak intensity or frequency. Thus, in considering (1) the auditory capability of a species, (2) its response to a particular sound, and (3) regulatory guidelines for exposure of animals to sound, the full dimensionality of sound should be taken into consideration. In this regard, factors that must be considered include the temporal and spectral characteristics of the sound, including the intensity, duration, duty cycle, frequency, bandwidth, rise time, temporal structure and similarity of any of these dimensions to biologically relevant sounds, as well as sensitivity of the relevant species to sounds with those characteristics.

Decisions based on such parameters should be made more objective by combining parameter values in mathematical decision models. Such risk assessment models have yet to be developed and tested for marine mammals and sound because of a lack of basic information about how sound characteristics are related to species-specific hearing sensitivity. An additional complication of modeling the effects of sound on marine mammals arises in predicting the received levels at the animals, particularly in shallow water, because of reflection off the sea surface and seafloor and unexpected areas of anomalous temperatures, salinities, and densities. Such factors can cause differences between predicted and actual received levels as great as 30 dB (J. Lynch, Woods Hole Oceanographic Institution, personal communication, 1999). This potential problem with acoustic propagation models creates a responsibility for major noise generators to calibrate their model-predicted sound levels against measured levels.

Recommendations: With regard to future research to study the effects of human-generated sound on marine mammals, the Committee supports the recommendation of NRC (1994) that there is a need for planned experiments designed to relate the behavior of specific animals to the received level of sound to which they are being exposed. Very few studies have succeeded in this aim. Because studies of ocean acoustics and marine mammal behavior are very challenging, successful experiments will require a closer collaboration between biologists and acousticians than has been the case in the past for many field studies. Success will also require continued refinement of techniques for making acoustic and visual observations, such as methods for locating vocalizing marine mammals and development of tags that can monitor received levels at the tagged animal.

To move beyond requiring extensive study of each sound source and each area in which it may be operated, a coordinated plan should be developed to explore how sound characteristics affect the responses of a representative set of marine mammal species in several biological contexts (e.g., feeding, migrating, and breeding). Research should be focused on studies of representative species using standard signal types, measuring a standard set of biological parameters, based on hearing type (Ketten, 1994), taxonomic group, and behavioral ecology (at least one species per group; Box 5.1). This could allow the development of mathematical models that predict the levels and types of noise that pose a risk of injury to marine mammals. Such models could be used to predict in multi-dimensional space where TTS is likely (a "TTS potential region") as a threshold of potential risk and to determine measures of behavioral disruption for different species groups. Observations should include both trained and wild animals. The results of such research could provide the necessary background for future environmental impact statements, regulations, and permitting processes.

The uncertainty in predictions of received sound levels hinders the application of models of marine mammal responses to sound and will require three complementary approaches: (1) development of better acoustic propagation models; (2) development of better observing systems to gather the data needed in models; and when the first two are not feasible, (3) development of better systems to observe ambient sound and transient noise pollution events in the ocean. Any research that includes relatively loud sound sources should monitor sound levels around the source site to gather data to calibrate its acoustic propagation models.

The locations of ATOC sites were dictated by requirements for the ATOC sound to reach many preplaced receivers and thus were not ideal for observing marine mammals. Pioneer Seamount is 88 km offshore, and observations on the north shore of Kauai were complicated by frequent high winds, waves, and bad weather. In general, studies designed to observe the effects of sound on marine mammals should be conducted in areas of high animal density for easy and less costly observation. Potential investigators should not transmit the noise until preexposure control data on individual subjects are obtained, and great care should be taken to ensure that ensonified marine mammals are not significantly damaged physiologically.

Acoustic studies focused on topics other than marine mammals should try to keep sound sources away from marine mammal "hotspots," even if this complicates logistics, increases costs, and/or decreases the efficiency of the experiments. In the case of the MMRP, acoustic and marine mammal studies with different goals were linked, leading to the proposal to place the sources in areas with high concentrations of marine mammals. Continuation of the oceanographic

BOX 5.1 Priority Research for Whales and Seals

Groupings of Species Estimated to Have Similar Sensitivity to Sound

Research and observations should be conducted on at least one species in each of the following seven groups:

1. Sperm whales (not to include other physterids)
2. Baleen whales
3. Beaked whales
4. Pygmy and dwarf sperm whales and porpoises (high-frequency [greater than 100 kHz] narrow-band sonar signals)
5. Delphinids (dolphins, white whales, narwhales, killer whales)
6. Phocids (true seals) and walruses
7. Otariids (eared seals and sea lions)

Signal Type

Standardized analytic signals should be developed for testing with individuals of the above seven species groups. These signals should emulate the signals used for human activities in the ocean, including impulse and continuous sources.

1. Impulse—air guns, explosions, sparkers, sonar pings.
2. Continuous—frequency-modulated (LFA and other sonars), amplitude-modulated (drilling rigs, animal sounds, ship engines), broadband (ship noise, sonar).

Biological Parameters to Measure

When testing representative species, several different biological parameters should be measured as a basis for future regulations and individual permitting decisions. These parameters include the following:

- Mortality
- TTS at signal frequency and other frequencies
- Injury—permanent threshold shifts
- Level B harassment
- Avoidance
- Masking (temporal and spectral)
- Absolute sensitivity
- Temporal integration function
- Nonauditory biological effects
- Biologically significant behaviors with the potential to change demographic parameters such as mortality and reproduction

component of ATOC would involve less risk to marine mammals if the source were moved, but this could be prohibitively expensive and would alter the data series. Even if scientists use sound responsibly, however, this does not guarantee the protection of essential marine mammal habitat, because commercial users are not subject to the same permitting requirements.

Studies of wild marine mammals should include careful determination of their locations, coupled with improved sampling and modeling of acoustic propagation to estimate received sound levels accurately. Alternatively, acoustic data loggers could be mounted on individual animals to record (1) the sounds (and their levels) to which the animals are exposed; (2) their vertical and horizontal movements; and (3) the sounds produced by the animals, including physiological sounds such as breathing and heartbeats. Preliminary analysis of MMRP data from tagged elephant seals demonstrated the usefulness of this approach in assessing behavioral response to ATOC sounds at Pioneer Seamount (Costa et al., 1999) and showed potential for use in other comparable studies. Acoustic data loggers will be particularly important for research on deep-diving mammals, whose behavior and exposure cannot be monitored systematically at depth in any other way.

A central theme of this report is that the task of developing predictive models of acoustic conditions that would harm marine mammals could be simplified by partitioning research among a small number of species that are representative in their hearing capabilities and sensitivities of larger groups of marine mammals. Box 5.1 describes the priority species groups, signal characteristics, and biological response parameters that should be investigated.

Richardson et al. (1995) summarized studies of marine mammal responses to human-generated sounds, particularly those associated with oil exploration and shipping. Some of these studies reported a significant difference between levels of pulsed versus more continuous sounds required to evoke a response in whales. To evoke the same level of response in migrating gray whales, a pulsed air gun sound required levels 50 dB higher than a diverse array of low-frequency continuous sources. This result is unexpected based on human hearing capabilities. How do marine mammals respond to signals with durations between the pulsed air gun noise (pulses separated by 7 to 15 seconds) and more continuous sounds? Another important question is: How do marine mammals respond when the received level is the same from two sources at different distances? This would help to discriminate whether marine mammals generally respond to received level (as was the case in the Phase II LFA study), estimated range to source, the gradient of acoustic energy over distance, and/or other sound characteristics.

Response to Stranded Marine Mammals

Findings: Observed behavioral responses of baleen whales to human-generated sound have all been reactions to sounds that are near or above the noise background (Richardson et al., 1995). Behavioral response thresholds in the range of 100 to 110 dB (Dahlheim and Ljungblad, 1990; Frankel et al., 1995) are considerably greater than human underwater threshold hearing levels. As has been suggested by Frankel et al. (1995) and Richardson et al. (1995), field observations of acoustic response thresholds probably have been limited by background noise rather than being indicative of true hearing thresholds.

The NRC (1994, p. 57) recommended the organization of a Stranded Whale Auditory Test (SWAT) team to obtain audiometric data from stranded or ensnared whales using recently developed electrophysiological techniques and instrumentation (e.g., Dolphin, 1997) for evoked potential studies (Hall, 1992). Some thresholds for killer whales and dolphins have been obtained using evoked potential methods (Popov and Supin, 1998; Popov and Klishin, 1998; Popov et al., 1998; Szymanski et al., 1995, 1998), but further research on methods of evoked potential audiometry are required for the application of the SWAT team approach to large whales.

Evoked potential audiograms, even on a few animals (e.g., using the SWAT team approach), would resolve the issue of whether baleen whale thresholds are below the observed behavioral response. Levels producing TTS often are stated as decibels above absolute sensitivity, so knowing the audiogram would be important for regulatory decisions and policy making if noise levels are being controlled based on TTS. Determining the hearing capabilities of the marine mammals in the categories listed in Box 5.1 is an exceptionally high priority for future research, and plans for such studies should be developed and implemented as soon as possible. The ultimate goal of such studies should be to provide information on hearing sensitivity, TTS, nonauditory effects, and other biological parameters listed in Box 5.1 to help in the determination of sound levels that might affect marine mammal hearing or significantly alter their behavior.

Physiological methods such as the auditory brainstem response (ABR) provide baseline data for use in evaluating promising behavioral techniques such as playbacks like those employed by Dahlheim and Ljungblad (1990) and others and thus are an important complement to behavioral techniques. So far these playback methods have produced thresholds that are on the order of 50 dB less sensitive than thresholds at the most sensitive frequencies obtained in the laboratory setting using ABR techniques with smaller species.

Recommendations: The concept of SWAT teams recommended in NRC (1994) and NRDC (1999) should be implemented by funding trained scientists and associating them with stranding networks. The Office of Naval Research (ONR) partially funded a small effort to support the activities of a SWAT team, but the hardware and field methods are not yet adequate for wide testing. The ONR program manager (R. Gisiner) estimates that a considerable, but not unreasonable, amount of hardware and software design and testing will be needed (about 1 to 2 years of effort) before a system capable of regular operation under the

SWAT team approach is feasible. However, this activity should be expanded to at least two teams, one on the east coast and one on the west coast of the United States. The teams should be responsible for (1) necropsy of suspected/possible marine mammal victims of sound injury (to be able to show whether sound caused the injuries or deaths) and (2) testing of hearing on stranded or entangled live animals. There is a need to expand the pool of individuals capable of doing this kind of work and capable of relating ear anatomy to function. An immediate need is for funding a specialist in evoked potential audiometry to develop improved methods applicable to large whales. A postdoctoral fellowship might be the most economical way to achieve this goal. NMFS and/or ONR should include funding for such work in the next budget cycle. Alternative possibilities for studying hearing in animals that are not kept in captivity also should be explored, such as placing a tag with electrodes on the head of a free-swimming whale and playing sound to the animal in a quiet environment.

Multiagency Research Support

Findings: Most marine mammal studies are funded from mission-oriented sources. At this time the greatest source of funding for marine mammal research is ONR. However, by its nature, ONR-funded research tends to be focused on questions of practical importance to the Navy and is not necessarily responsive to the broad interests of scientists seeking to learn more about the basic biology of marine mammals. Scientist-driven fundamental research could significantly improve our understanding of hearing and the effects of low-frequency sound on marine mammals, as well as our overall understanding of the acoustic behavior of these animals.

Recommendations: If government funding shortages and priorities continue to constrain budgets for marine mammal research in the foreseeable future, management of sound in the ocean should remain conservative (and should incorporate management of all sources of human-generated noise in the sea, including industrial sources), in the absence of required knowledge. If government regulators need better information on which to base decisions, they should take such steps as necessary to provide increased funding for marine mammal research and to improve the ways that needed research is identified, funded, and conducted. Acquiring better information is often complicated because the regulatory parts of agencies like NMFS and FWS are separated from research, and funded research may not necessarily match research needed by regulators. It is imperative that the research and regulatory arms of NMFS and FWS maintain good linkages within these two agencies, and that priority is given to research needed by regulators in each agency. Government agencies with basic science missions (e.g., National Science Foundation [NSF] and National Institutes of Health [NIH]) should fund marine mammal research at the levels needed to answer fundamental questions about hearing anatomy and physiology. Mission agencies with responsibilities related to marine mammals (e.g., ONR, National Oceanic and Atmospheric Administration [NOAA], MMS, U.S. Geological Survey [USGS]) should also fund basic research (notwithstanding ONR's limitations under the Mansfield Amendment), in the spirit of the recommendation of NRC (1992) that "federal agencies with marine-related missions find mechanisms to guarantee the continuing vitality of the underlying basic science on which they depend" (p. 28). Such research should receive the same level of peer review as other basic research and be competitive with such programs for funding. Because marine mammal research is quite expensive, multiagency funding may be necessary to spread the costs. Alternatively, multiple parts of the same agency may need to cooperate in order to provide sufficient funds.

Multidisciplinary Research Teams and Peer Review

Findings: The MMRP suffered from the lack of an interdisciplinary group for planning its research. Further, the MMRP was added to an existing research program, rather than being designed to fulfill its own objectives. The agenda for the acoustic oceanographic component of ATOC required different transmission schedules from those that would have been optimal for marine mammal research. For example, the geographic location, depth, and duty cycle (in part) of the source were determined by the needs of ATOC, not the needs of the MMRP. As a consequence, the biological data that resulted were not optimal for answering the fundamental biological questions raised by the ATOC transmissions.

Recommendation: Consideration should be given to establishing a multi-investigator program to study the effects of sound on marine mammals, funded by consortia of government agencies, non-governmental organizations, shipping, and hydrocarbon exploration and production industries. These consortia should include individuals, organizations, and companies in nations that share marine mammal stocks and sound-producing activities with the United States (e.g., Canada, Mexico, nations of the North Atlantic Treaty Organization). Such consortia could be initiated through a workshop to bring together the interested communities. The design and implementation of auditory research on marine mammals ideally should be an interdisciplinary enterprise. Valuable contributions can be made by physical acousticians on the choice of sound stimuli to be used, by electronics experts on the choice and calibration of transducers for presenting the stimuli, by marine biologists on the choice of species and/or the best season and location for testing, by psychoacousticians on the testing procedures, and by statisticians on initial design and eventual data analysis and presentation. Without collaboration among specialists within these various disciplines, there is a greater probability that expensive and time-consuming projects will contain errors that preclude an unambiguous interpretation of the results. These projects are sufficiently complex that one or two individuals cannot reasonably be expected to have the full range of knowledge necessary for success. The logistical difficulties, permitting issues, and expense of such research demand advanced planning in all these areas.

If such a research program is established, it should use a public Request for Proposal (RFP) process that results in proposals from more than one research team and is modeled after the peer-review processes used by NSF and NIH. Conversely, some research should continue to be funded through the less conservative ONR model, which provides program managers with greater latitude to fund more innovative science. A spectrum of funding styles is useful. The RFP should be well advertised to encourage ideas and proposals from a wide range of researchers and institutions (including foreign participants), rather than relying on a set of traditional investigators. The goal of the process should be to optimize the selection of hypotheses, methods, and design and to identify the best performer(s) (e.g., best track record in previous work) for the proposed work. It is to the advantage of the sponsors to implement programs of broad-based peer review for such proposals. Future research on marine mammals unquestionably would profit from a broad-based review of the plans developed by multidisciplinary teams and evaluated by a peer-review process that is objective and independent. Such a review should determine whether the proposing investigative teams did the following adequately:

- identified basic problem(s);
- established specific hypotheses to be tested, with appropriate methods for data reduction, data presentation, and statistical analysis;
- identified optimal experimental methods and test conditions (including geographic location of study); and
- evaluated the power of the proposed experimental design.

Because long research projects often need to adjust to experience gained in field programs and learning about what kinds of observations are practical and achievable, it is important to provide advice from an outside review team later in the life of a project.

Sponsors of research need to be aware that studies funded and led by one special interest are vulnerable to concerns about conflict of interest. For example, research on the effects of smoking funded by NIH is likely to be perceived to be more objective than research conducted by the tobacco industry. Concern for peer review, efficiency, and independence argues for having an agency such as NSF take the lead in managing an interagency research program on the effects of noise on marine mammals.

Agencies that fund such applied research should ensure that adequate funding for analysis and plans for peer review are in place before a research award is made. Analysis might be speeded by employing a larger team for analysis and involving this team in planning the observations to make them as easy as possible to analyze later. Although publication in peer-reviewed journals is the standard by which most research is judged, applied research output from projects like the MMRP is not necessarily suitable for publication in available academic journals and the results may need to be used for regulatory decisions within a shorter amount of time than the normal journal paper cycle. Timely peer review of such studies might be better accomplished by conducting a mail and/or panel review of results by an independent group established specifically for this purpose.

Population-Level Audiograms

Findings: Ridgway and Carder (1997) published the first evidence that auditory capabilities in bottlenose dolphins may vary with sex and may change with age, similar to observations in humans (Ward, 1997). These data reinforce the recommendation of NRC (1994) that audiograms should be obtained for many individuals in a population to determine the normal range of hearing capability and the effects of aging. Because of the difficulty and expense of training and maintaining large numbers of animals, most studies collect data from one or two individuals of a particular species. Although individual differences have been noted (e.g., Terhune and Turnbull, 1995; Schlundt et al., 2000), no single study has used the same methods for multiple individuals of both sexes and varied ages. In addition, Ridgway and Carder (1997) reported that a young dolphin apparently had been deafened due to disease and had survived in the wild although deaf and mute. Clearly, there is a range of normal hearing among individuals, and even deficits may not prove fatal for individuals of social species. The major barrier to large-scale testing of the hearing of many individuals of the same species has been the need to train each individual to respond to sounds in measurable ways. The further development of audiometric procedures based on auditory evoked potentials (Hall, 1992; Szymanski et al., 1999) would eliminate that problem.

Recommendations: Federal agencies should sponsor studies on the hearing abilities of both free-swimming and stranded animals. Population-level audiograms of many individuals (such as are performed for humans; see Yost and Killion, 1997) are necessary for establishing the baseline of hearing capabilities and normal hearing loss in marine mammals, as also recommended in NRC (1994). Stranded animals should be assessed to determine if their hearing is "normal." Data are needed to provide comparisons that would allow an evaluation of how common hearing deficits may be among stranded animals. The development of population-level audiograms will require the perfection and wide use of auditory evoked potential techniques, to eliminate the need to train all tested animals. However, if the cost and techniques limit widespread auditory evoked potential measurements of captive animals, a good sample of multiple animals (different ages and both genders) of the same species should be tested.

National Captive Marine Mammal Research Facility

Findings: There are few sources of trained marine mammals and few facilities available to academic (or even government) scientists for closely controlled research on the hearing capabilities of captive marine mammals and on sick or injured marine mammals being rehabilitated for release back into the wild. The costs of capturing, training, and maintaining marine mammals are great, meaning that anyone working with a marine mammal must make a long-term commitment to its care and well-being. Unlike work with small lab mammals or farm animals, marine mammal research requires decades of obligation to the animals, including experienced trainers and veterinarians and long-term care of trained and retired animals. Some rare marine mammals will only become available for study at such facilities that have received them after rescues. Lack of specialized research facilities hinders the priority studies described earlier. For example, a rare opportunity to obtain evoked potential audiometric data from a stranded baleen whale (a juvenile gray whale) was lost when competing demands for access made it impossible to get sufficient time with the animal to test its hearing. This experience emphasizes the need for not only better methods of obtaining auditory information, but also for gaining access to species for which few or no auditory data exist.

Currently, there is only one site in the United States (and perhaps the world) that has the facilities and animals that could be used in such studies. This site is operated by the U.S. Navy in San Diego, California.[1] Even this site has its limitations, however, in that investigators must be U.S. citizens to work with the trained animals. Although some research on the hearing abilities of marine mammals could possibly be conducted at public aquariums, research commitments at aquariums usually are secondary to public display requirements. In addition, they are not able to do research with animals in the open sea, as is possible with the highly trained animals maintained by the Navy. Such training takes years and is beyond the capabilities or interests of public aquariums. The ocean science community has a variety of different centers and shared-used facilities, for example, the Ocean Drilling Program, that could provide a model for a national captive marine mammal research facility.

Recommendations: If the studies described in Chapter 3 and Box 5.1 are of sufficient priority to reduce uncertainties in the regulation of human-generated sound in the ocean, federal agencies should establish a national facility for the study of marine mammal hearing and behavior. The Committee believes that such a facility might be established at relatively little incremental cost by enhancement of the existing Navy facility.

---

[1]The one committee member associated with this facility did not take a position on whether the facility could or should be expanded and made available beyond Navy scientists.

The facility for captive marine mammal research would have animals for "hire" by investigators funded for peer-reviewed research. Offset funds would come from individual grants and researchers, but the funding base for such a facility should not be provided solely by such offsets. Allocation of space, animals, and facility resources should be determined by a broad-based review board on the basis of the quality and significance of the proposed research. An additional virtue of establishing a national captive marine mammal research facility is that the total number of marine mammals removed from the wild would be minimized. Investigators could apply for support for short- or long-term study of the animals at this facility, from the range of agencies funding marine mammal research, at costs that would not have to include long-term maintenance of the animals. Such a facility should include the capability to work with trained animals in the open ocean. The Navy's Marine Mammal Program facility in San Diego keeps marine mammals and already has trained animals and expertise in maintaining them. Its role potentially could be expanded to provide a more widely accessible national facility, including unclassified research. If such a facility is operated by the Navy, it will be necessary to ensure that research data are not restricted from publication. Establishment of a facility to promote field studies could also enable research recommended in this report, but such a facility would be more expensive and a lower priority than a national facility for research on trained, captive animals.

REGULATORY REFORM

Findings: The existing permit system for acoustic research is unnecessarily restrictive in some aspects and not comprehensive enough in other regards. It is not scientifically defensible to apply general source standards for permit requirements (120 dB for continuous sound, 160 dB for intermittent sound, and 180 dB for sounds of all frequencies and durations) for all species and all sound characteristics under all possible conditions. Until NMFS publishes new acoustic guidelines, current NMFS policy recommends applying for a small-take authorization for sound-producing research activities that have the potential to harass, injure, or kill a marine mammal (K. Hollingshead, NMFS, personal communication, 1999). Different species have different sensitivities and susceptibilities, and sound effects may accumulate as new sources are added. In the absence of information, managers rightly have chosen to be cautious, at least in regard to permitting ocean science research. However, as noted in NRDC (1999), there is virtually no regulation of sound produced by large commercial ocean-going vessels: "The worst polluter, shipping, is also the least regulated, while a comparative lightweight [in terms of the amount of sound put into the ocean], scientific research, is far more strictly scrutinized" (p. 13). This combination of cautious regulation of minor activities, coupled with lack of regulation of major noise sources, will not, in the long run, adequately protect marine mammals from potentially deleterious effects of noise and could unnecessarily impede important acoustic research.

Recommendations: Congress should change the Marine Mammal Protection Act (MMPA) and/or NOAA should change the implementing legislation of the MMPA to allow incidental take authorization based solely on negligible impact on the population. Research should be undertaken to allow the definition of Level A harassment to be related to the TTS produced in a species, when known. Level B harassment should be limited to meaningful disruption of biologically significant activities that could affect demographically important variables such as reproduction and longevity.

COMPREHENSIVE MONITORING AND REGULATION OF SOUND IN THE OCEAN

Findings: Protecting marine mammals from significant adverse impacts clearly requires a broader application of regulations. There is a global increase of sound levels in the sea resulting from shipping, recreation, aircraft, and naval operations as well as research (Urick, 1986). It is important to characterize the existing ambient sound field in terms of the levels, frequencies, and locations of sources, especially in terms of areas particularly important for marine mammals (i.e., the "hotspots" of NRDC, 1999). Such a characterization of the ambient noise field will provide a context for determining when, where, and with what characteristics new sources could be added.

Recommendations: Noise monitoring is important and acoustic hotspots should be identified. Fortunately, ambient noise data exist for a variety of locations, which could provide time series and baselines for additional monitoring. Existing data should be identified and made accessible through a single easy-to-access source. Like marine mammal research programs, funding for noise monitoring should be awarded based on responses to a request for proposals and careful evaluation of the costs and benefits of the proposed systems. The opening of the existing IUSS for whale research was important for demonstrating the power of bottom-mounted hydrophone arrays, but the IUSS may or may not provide the best system for the acoustic monitoring tasks envisioned here, given that it was designed for an entirely different purpose.

The first step in comprehensive monitoring and regulation of sound in the ocean should be to attempt to characterize the existing ambient sound field in the ocean and to characterize the sources that contribute to it. Monitoring of baseline sound levels should be carried out, particularly in critical habitats of acoustically sensitive or vulnerable species or in habitats critical to specific life stages, such as breeding and calving areas. Protection of marine mammals from subtle or long-term effects of harassment cannot be achieved through regulation of individual "takes." An alternative habitat-oriented approach is required to protect marine mammals from the cumulative impacts of noise pollution, chemical pollution, physical habitat loss, and fishing. Such an approach requires monitoring of the status of marine mammal populations along with the quality of critical habitats, including the acoustic quality. Account should be taken of the populations involved; it is sensible to protect more rigorously species that are more endangered (e.g., northern right whales) than those that are less at risk. Basic research regarding what is significant about critical habitats and what factors have population-level effects—for example, food supply, water quality, and noise levels and characteristics—will prove much more effective for protecting marine mammals than merely attempting to regulate individual human activities that may potentially cause changes in the behavior of an individual marine mammal. NMFS regulations should encompass the entirety of noise pollution and other threats to marine mammals.

References

Aburto, A., D.J. Rountry, and J.L. Danzer. 1997. *Behavioral Response of Blue Whales to Active Signals.* Technical Report 1746. National Technical Information Service, Springfield, VA.

Advanced Research Projects Agency (ARPA). 1995. *Final Environmental Impact Statement for the California Acoustic Thermometry of Ocean Climate Project and Its Associated Marine Mammal Research Program.* Marine Acoustics, Inc., Arlington, VA.

Advanced Research Projects Agency (ARPA) and National Oceanic and Atmospheric Administration (NOAA). 1995. *Final Environmental Impact Statement for the Kauai Acoustic Thermometry of Ocean Climate Project and Its Associated Marine Mammal Research Program.* Marine Acoustics, Inc., Arlington, VA.

Ahroon, W.A., R.P. Hamernik, and S.F. Lei. 1996. The effects of reverberant blast waves on the auditory system. *Journal of the Acoustical Society of America* 1009:2247-2257.

Andre, M., M. Terada, and Y. Watanabe. 1997. Sperm whale (*Physeter macrocephalus*) behavioral response after the playback of artificial sounds. *Reports of the International Whaling Commission* 47:499-504.

ATOC Consortium. 1998. Ocean climate change: Comparison of acoustic tomography, satellite altimetry, and modeling. *Science* 281:1327-1332.

Au, W.L.L., P.E. Nachtigall, and J.L. Pawloski. 1997. Acoustic effects of the ATOC signal (75 Hz, 195 dB) on dolphins and whales. *Journal of the Acoustical Society of America* 101:2973-2977.

Baggeroer, A., and W. Munk. 1992. The Heard Island Feasibility Test. *Physics Today* 45:22-30.

Banner, A., and M. Hyatt. 1973. Effects of noise on eggs and larvae of two estuarine fish. *Transactions of the American Fisheries Society* 102:134-136.

Bowles, A.E., M. Smultea, B. Würsig, D.P. DeMaster, and D. Palka. 1994. Relative abundance and behavior of marine mammals exposed to transmissions from the Heard Island Feasibility Test. *Journal of the Acoustical Society of America* 96:2469-2484.

Burgess, W.C., P.L. Tyack, B.J. LeBoeuf, and D.P. Costa. 1998. A programmable acoustic recording tag and first results from free-ranging northern elephant seals. *Deep-Sea Research* 45:1327-1351.

Calambokidis, J. 1996. *Preliminary (Quick-Look) Analysis of Aerial Survey Data Surveys Conducted Through March 1996.* ATOC Program Office, Scripps Institution of Oceanography, La Jolla, CA.

Calambokidis, J. 1999. Aerial surveys conducted off Pioneer Seamount for the ATOC Marine Mammal Research Program. Unpublished report to MMRP Advisory Board.

Clark, C.W. 1995. Application of U.S. Navy underwater hydrophone arrays for scientific research on whales. *Reports of the International Whaling Commission* 45:210-212.

Clark, C.W., and J.M. Clark. 1980. Sound playback experiments with southern right whales. *Science* 207:663-665.

Corwin, J.T., and J.C. Oberholtzer. 1997. Fish 'n chicks: Model recipes for hair-cell regeneration? *Neuron* 19:951-954.

Costa, D.P., D.E. Crocker, J. Gedamke, P.M. Webb, D. Houser, and S. Blackwell. 1999. Effects of the ATOC sound source on the diving behavior of northern elephant seals, *Mirounga angustirostris*. Unpublished paper, University of California, Santa Cruz.

Costa, D.P., and T.M. Williams. 1999. Marine mammal energetics. Pp. 176-217 in *Biology of Marine Mammals*, J.E. Reynolds III and John R. Twiss, Jr. (eds.), Smithsonian Press, Washington, DC.

Crane, N.L., and K. Lashkari. 1996. Sound production of gray whales, *Eschrichtius robustus*, along their migration route: A new approach to signal analysis. *Journal of the Acoustical Society of America* 100:1878-1886.

Crum, L.A., and Y. Mao. 1996. Acoustically enhanced bubble growth at low frequencies and its implications for human diver and marine mammal safety. *Journal of the Acoustical Society of America* 99:2898-2907.

Dahlheim, M.E., and D.K. Ljungblad. 1990. Preliminary hearing study on gray whales (*Eschrichtius robustus*) in the field. Pp. 335-346 in *Sensory Abilities of Cetaceans: Laboratory and Field Evidence*, J.A. Thomas and R.A. Kastelein (eds.). Plenum Press, New York.

D'Amico, A. 1998. Summary Record, SACLANTCEN Bioacoustics Panel, La Spezia, Italy, June 15-17, 1998 (available at *http://www.saclantc.nato.int/whales/*).

Decory, L., A.L. Dancer, and J-M Aran. 1992. Species differences and mechanisms of damage. Pp. 73-88 in *Noise-Induced Hearing Loss,* A.L. Dancer, D. Henderson, R.J. Salvi, and R.P. Hamernik (eds.). Mosby, St. Louis, MO.

Demski, L., G.W. Gerald, and A.N. Popper. 1973. Central and peripheral mechanisms in teleost sound production. *American Zoologist* 13:1141-1167.

Department of Commerce (DOC). 1995. Small takes of marine mammals; harassment takings incidental to specified activities. *Federal Register* 60(104):28379-28386.

Department of Commerce (DOC). 1996. Small takes of marine mammals; harassment takings incidental to specified activities in arctic waters; regulation consolidation; update of Office of Management and Budget (OMB) approval numbers. *Federal Register* 61 (70):15884-15891.

Dolphin, W.F. 1995. The envelope following response in three species of cetaceans. Pp. 147-172 in *Sensory Systems of Aquatic Mammals*, R.A. Kastelein, J.A. Thomas, and P.E. Nachtigall (eds.). De Spil Publishers, Woerden, The Netherlands.

Dolphin, W.F. 1996. Auditory evoked responses to amplitude modulated stimuli consisting of multiple envelope components. *Journal of Comparative Physiology A* 179:113-121.

Dolphin, W.F. 1997. Electrophysiological measures of auditory processing in odontocetes. *Bioacoustics* 8:79-101.

Dolphin, W.F., W.W.L. Au, P.E. Nachtigall, and J. Pawloski. 1995. Modulation rate transfer functions to low-frequency carriers in three species of cetaceans. *Journal of Comparative Physiology A* 177:235-245.

Edds, P.L. 1982. Vocalizations of the blue whale, *Balaenoptera musculus*, in the St. Lawrence River. *Journal of Mammalogy* 63:345-347.

Edds, P.L. 1988. Characteristics of finback *Balaenoptera physalus* vocalizations in the St. Lawrence Estuary. *Bioacoustics* 1(2/3):131-149.

Edds-Walton, P.L. 1997. Acoustic communication signals of mysticete whales. *Bioacoustics* 8(1&2):47-60.

Edds-Walton, P.L. 2000. Vocalizations of minke whales (*Balaenoptera acutorostrata*) in the St. Lawrence Estuary. *Bioacoustics*, in press.

Erbe, C., and D.M. Farmer. 1998. Masked hearing thresholds of a beluga whale (*Delphinapterus leucas*) in icebreaker noise. *Deep-Sea Research II* 45:1373-1388.

Fluur, E., and J. Adolfson. 1966. Hearing in hyperbaric air. *Aerospace Medicine* 37:783-785.

Frankel, A.S., and C.W. Clark. 1998a. Results of low-frequency playback of M-sequence noise to humpback whales, *Megaptera novaeangliae*, in Hawaii. *Canadian Journal of Zoology* 76:521-535.

Frankel, A., and C.W. Clark. 1998b. Acoustic Thermometry of Ocean Climate: Quick-Look Report of the Hawai'i ATOC-MMRP Hawaiian 1997/98 Results. Unpublished paper.

Frankel, A., and C.W. Clark. 1999a. Behavioral responses of humpback whales to operational ATOC signals. Unpublished paper, Cornell University.

Frankel, A., and C.W. Clark. 1999b. Factors affecting the distribution and abundance of humpback whales off the north shore of Kaua'i. Unpublished paper, Cornell University.

Frankel, A.S., J.R. Mobley, Jr., and L.M. Herman. 1995. Estimation of auditory response thresholds in humpback whales using biologically meaningful sounds. Pp. 55-70 in *Sensory Systems of Aquatic Mammals*, R.A. Kastelein, J.A. Thomas, and P.E. Nachtigal (eds.). De Spil Publishers, Woerden, The Netherlands.

Frantzis, A. 1998. Does acoustic testing strand whales? *Nature* 392:29.

Gisiner, R. (ed.). 1998. *Proceedings. Workshop on the Effects of Anthropogenic Noise in the Marine Environment.* Office of Naval Research, Arlington, VA.

Gordon, J.C.D., D. Gillespie, L.E. Rendell, and R. Leaper. 1996. Draft report on playback of ATOC-like sounds to sperm whales (*Physeter macrocephalus*) off the Azores. Unpublished manuscript submitted to the ATOC Marine Mammal Research Program, Bioacoustics Research Program, Laboratory of Ornithology, Cornell University, Ithaca, NY.

Hall, J.W., III. 1992. *Handbook of Auditory Evoked Responses.* Allyn and Bacon, Boston.

Harvey, J.T., and T. Eguchi. 1997. Effects of ATOC sounds on the harbor seal, *Phoca vitulina Richardsi*, in Monterey Bay. National Technical Information Service, Springfield, VA.

Hastings, M.C., A.N. Popper, J.J. Finneran, and P.J. Lanford. 1996. Effect of low-frequency underwater sound on hair cells of the inner ear and lateral line of the teleost fish, *Astronotus ocellatus*. *Journal of the Acoustical Society of America* 99:1759-1766.

Hunter-Duvar, I.M., and G. Bredberg. 1974. Effects of intense auditory stimulation: Hearing losses and inner ear changes in the chinchilla. *Journal of the Acoustical Society of America* 55:795-801.

Joint Oceanographic Institutions (JOI). 1994. *Dual Use of IUSS: Telescopes in the Ocean.* Joint Oceanographic Institutions, Inc., Washington, DC.

Kastak, D., and R.J. Schusterman. 1996. Temporary threshold shift in a harbor seal (*Phoca vitulina*). *Journal of the Acoustical Society of America* 100:1905-1908.

Kastak, D., and R.J. Schusterman. 1998. Low-frequency amphibious hearing in pinnipeds: Methods, measurements, noise, and ecology. *Journal of the Acoustical Society of America* 103:2216-2228.

Kastak, D., R.J. Schusterman, B.L. Southall, and C.J. Reichmuth. 1999. Underwater temporary threshold shift induced by octave band noise in three species of pinniped. *Journal of the Acoustical Society of America* 106:1142-1148.

Ketten, D.R. 1994. Functional analyses of whale ears: Adaptations for underwater hearing. *Institute of Electrical and Electronics Engineers Proceedings in Underwater Acoustics* 1:264-270.

Ketten, D.R. 1997. Structure and function in whale ears. *Bioacoustics* 8(1&2):103-136.

Ketten, D.R., J. Lien, and S. Todd. 1993. Blast injury in humpback whale ears: Evidence and implications. *Journal of the Acoustical Society of America* 94:1849-1850.

Klimley, A.P., and S.C. Beavers. 1998. Playback of acoustic thermometry of ocean climate (ATOC)-like signal to bony fishes to evaluate phonotaxis. *Journal of the Acoustical Society of America* 104:2506-2510.

Knudsen, V.O., R.S. Alford, and J.W. Emling. 1948. Underwater ambient noise. *Journal of Marine Research* 7(3):410-429.

Kryter, K.D. 1985. *The Effects of Noise on Man.* 2nd Edition. Academic Press, Orlando, FL.

Lagardére, J.P. 1982. Effects of noise on growth and reproduction of *Crangon crangon* in rearing tanks. *Marine Biology* 71:177-186.

Lapedes, D.N. (Editor-in-chief). 1974. *McGraw-Hill Dictionary of Scientific and Technical Terms.* McGraw-Hill, New York.

Lesage, V., C. Barrette, M.C.S. Kingsley, and B. Sjare. 1999. The effect of vessel noise on the vocal behavior of belugas in the St. Lawrence Estuary, Canada. *Marine Mammal Science* 15(1):65-84.

Lettvin, J.Y., E.R. Grumberg, R.M. Rose, and G. Plotkin. 1982. Dolphins and the bends. *Science* 216:650-651.

Levendag, P.C., W. Kuijpers, J.J. Eggermont, P. van den Brock, H. Huibers, and P.L. Huygen. 1981. The inner ear and hyperbaric conditions: An electrophysiological and morphological study. *Acta Oto-Laryngologica Supplement* 382:1-110.

Liberman, M.C., and L.W. Dodds. 1987. Acute ultrastructural changes in acoustic trauma: Serial-section reconstruction of stereocilia and cuticular plates. *Hearing Research* 26:45-64.

Ljungblad, D.K., K.M. Stafford, H. Shimada, and K. Matsuoka. 1997. Sounds attributed to blue whales recorded off the southwest coast of Australia in December 1995. *Reports of the International Whaling Commission* 47:435-439.

Lombarte, A., H.Y. Yan, A.N. Popper, J.C. Chang, and C. Platt. 1993. Damage and regeneration of hair cell ciliary bundles in a fish ear following treatment with gentamicin. *Hearing Research* 66:166-174.

Lonsbury-Martin, B.L., G.K. Martin, and B.A. Bohne. 1987. Repeated TTS exposures in monkeys: Alterations in hearing, cochlear structure, and single-unit thresholds. *Journal of the Acoustical Society of America* 81:1507-1518.

Lorenz, K. 1981. *The Foundations of Ethology.* Springer-Verlag, NY.

Luz, G.A., and D.M. Lipscomb. 1973. Susceptibility to damage from impulse noise: Chinchilla versus man or monkey. *Journal of the Acoustical Society of America* 54:1750-1754.

Malme, C.I., P.R. Miles, C.W. Clark, P. Tyack, and J.E. Bird. 1983. Investigations of the potential effects of underwater noise from petroleum industry activities on migrating gray whale behavior. *Bolt Beranek and Newman Report No. 5366* submitted to the Minerals Management Service, U.S. Department of the Interior, Washington, DC. NTIS PB86-218377.

Malme, C.I., P.R. Miles, C.W. Clark, P. Tyack, and J.E. Bird. 1984. Investigations of the potential effects of underwater noise from petroleum industry activities on migrating gray whale behavior. Phase II. January 1984 migration. *Bolt Beranek and Newman Report No. 5586* submitted to the Minerals Management Service, U.S. Department of the Interior, Washington, DC. NTIS PB86-218377.

Malme, C.I., P.R. Miles, G.S. Miller, W.J. Richardson, and D.G. Roseneau. 1989. *Analysis and Ranking of the Acoustic Disturbance Potential of Petroleum Industry Activities and Other Sources of Noise in the Environment of Marine Mammals in Alaska.* Mineral Management Service, Anchorage.

Mate, B.R, K.M. Stafford, and D.K. Ljungblad. 1994. A change in sperm whale (*Physeter macrocephalus*) distribution correlated to seismic surveys in the Gulf of Mexico. *Journal of the Acoustical Society of America* 96(5:Part 2):3268-3269.

Mattlin, R.H. 1995. *Effect of Low-Frequency Sound on Seasonal Foraging Ecology and Diving Behavior of the New Zealand Fur Seal.* National Technical Information Service, Springfield, VA.

McDonald, M.A., J.A. Hildebrand, and S.C. Webb. 1995. Blue and fin whales observed on a seafloor array in the Northeast Pacific. *Journal of the Acoustical Society of America* 98:712-721.

McFadden, D. 1986. The curious half-octave shift: Evidence of a basalward migration of the traveling-wave envelope with increasing intensity. Pp. 295-312 in *Applied and Basic Aspects of Noise-Induced Hearing Loss*, R. Salvi, D. Henderson, R.P. Hamernik, and V. Coletti (eds.). Plenum Press, New York.

Miller, J.D. 1974. Effects of noise on people. *Journal of the Acoustical Society of America* 56: 729-764.

Mobley, J.R., L.M. Herman, and A.S. Frankel. 1988. Responses of wintering humpback whales (*Megaptera novaeangliae*) to playback of recordings of winter and summer vocalizations and of synthetic sound. *Behavioral Ecology and Sociobiology* 23:211-223.

Mobley, J.R., Jr., R.A. Grotefendt, P.H. Forestell, and A. Frankel. 1999. Results of aerial surveys of marine mammals in the major Hawaiian islands (1993-1998). Report to the Acoustic Thermometry of Ocean Climate Marine Mammal Research Program (ATOC MMRP). Unpublished paper.

Moore, K.E., W.A. Watkins, and P.L. Tyack. 1993. Pattern similarity in shared codas from sperm whales (*Physeter catodon*). *Marine Mammal Science* 9:1-9.

Moore, S.E., M.E. Dahlheim, K.M. Stafford, C.G. Fox, H.W. Braham, M.A. McDonald, and J. Thomason. 1999. Acoustic and visual detection of large whales in the eastern north Pacific Ocean. U.S. Department of Commerce, NOAA Technical Memorandum NMFS-AFSC-107.

Munk, W., and C. Wunsch. 1979. Ocean acoustic tomography: A scheme for large-scale monitoring. *Deep-Sea Research* 26A:439-464.

Munk, W., P. Worcester, and C. Wunsch. 1995. *Ocean Acoustic Tomography*. Cambridge University Press, Cambridge.

Myrberg, A.A., Jr. 1972. Using sound to influence the behaviour of free-ranging marine animals. Pp. 435-468 in *Behaviour of Marine Animals*, Vol. 2, H.E. Winn and B.L. Olla (eds). Plenum Press, New York.

Myrberg, A.A., Jr. 1978. Underwater sound: Its effect on the behavior of sharks. Pp. 391-417 in *Sensory Biology of Sharks, Skates and Rays*, E.S. Hodgson and R.F. Mathewson (eds.). Government Printing Office, Washington, DC.

Myrberg, A.A., Jr. 1980. Ocean noise and the behavior of marine animals. Pp. 461-491 in *Advanced Concepts in Ocean Measurements for Marine Biology*, F.P. Diemer, F.J. Vernberg, and D.V. Mirkes (eds.). University of South Carolina Press, Columbia.

Myrberg, A.A., Jr. 1990. The effects of man-made noise on the behavior of marine animals. *Environment International* 16:575-586.

Myrberg, A.A., Jr., C.R. Gordon, and A.P. Klimley. 1976. Attraction of free ranging sharks by low-frequency sound, with comments on its biological significance. Pp. 205-228 in *Sound Reception in Fish*, A. Schuijf and A.D. Hawkins (eds.). Elsevier, Amsterdam.

National Marine Fisheries Service (NMFS). 1996. *Our Living Oceans. Report on the Status of U.S. Living Marine Resources*. National Oceanic and Atmospheric Administration, Silver Spring, MD.

National Research Council (NRC). 1992. *Oceanography in the Next Decade: Building New Partnerships*. National Academy Press, Washington, DC.

National Research Council (NRC). 1994. *Low-Frequency Sound and Marine Mammals: Current Knowledge and Research Needs*. National Academy Press, Washington, DC.

National Research Council (NRC). 1996. *Marine Mammals and Low-Frequency Sound: Progress Since 1994–An Interim Report*. National Academy Press, Washington, DC.

National Research Council (NRC). 1999. *Building Ocean Science Partnerships: The United States and Mexico Working Together*. National Academy Press, Washington, DC.

National Research Council (NRC). 2000. *Reconciling Observations of Global Temperature Change.* National Academy Press, Washington, DC.

Natural Resources Defense Council (NRDC). 1999. *Sounding the Depths: Supertankers, Sonar, and the Rise of Undersea Noise.* Natural Resources Defense Council, Inc., New York.

Nixon, J.C., and A. Glorig. 1961. Noise-induced permanent threshold shift at 2000 cps and 4000 cps. *Journal of the Acoustical Society of America* 33:904-908.

Pantev, M., and Ch. Pantev. 1979. Cortical auditory evoked responses under hyperbaric conditions. Pp. 241-244 in *Models of the Auditory System and Related Signal Processing Techniques*, M. Hoke and E. de Boer (eds.). *Scandinavian Audiology Supplement* 9.

Plomp, R., and M.A. Bouman. 1959. Relation between hearing threshold and duration for tone pulses. *Journal of the Acoustical Society of America* 31:749-758.

Popov, V.V., and V.O. Klishin. 1998. EEG study of hearing in the common dolphin. *Aquatic Mammals* 24:13-20.

Popov, V.V., and A.Y. Supin. 1998. Auditory evoked responses to rhythmic sound pulses in dolphins. *Journal of Comparative Physiology A. Sensory Neural and Behavioral Physiology* 183:519-524.

Popov, V.V., A.Y. Supin, and V.O. Klishin. 1998. Frequency tuning of the dolphin's hearing as revealed by auditory brain-stem response with notch-noise masking. *Journal of the Acoustical Society of America* 102:3795-3801.

Reeves, R.R., R.J. Hofman, G.K. Silber, and D. Wilkinson (eds.). 1996. *Acoustic Deterrence of Harmful Marine Mammal-Fishery Interactions.* Marine Mammal Commission, Washington, DC.

Richardson, W.J., B. Würsig, and C.R. Greene, Jr. 1986. Reactions of bowhead whales, *Balaena mysticetus*, to seismic exploration in the Canadian Beaufort Sea. *Journal of the Acoustical Society of America* 79:1117-1128.

Richardson, W.J., C.R. Greene, Jr., C.I. Malme, and D.H. Thomson. 1995. *Marine Mammals and Noise.* Academic Press, New York.

Ridgway, S.H., and W.W.L. Au. 1999. Dolphin hearing and echolocation: The bottlenose dolphin, *Tursiops truncatus*. Pp. 858-862 in *Encyclopedia of Neuroscience*, 2nd Edition, G. Adelman and B. Smith (eds.). Springer-Verlag, New York.

Ridgway, S.H., and D.A. Carder. 1997. Hearing deficits measured in some *Tursiops truncatus*, and discovery of a deaf/mute dolphin. *Journal of the Acoustical Society of America* 101:590-594.

Ridgway, S.H., and R. Howard. 1979. Dolphin lung collapse and intramuscular circulation during free diving: Evidence from nitrogen washout. *Science* 206:1182-1183.

Ridgway, S.H., and R. Howard. 1982. Dolphins and the bends. *Science* 216:651.

Ridgway, S.H., D. Carder, R. Smith, T. Kamolnick, and W. Elsberry. 1997. First audiogram for marine mammals in the open sea: Hearing and whistling by two white whales down to 30 atmospheres. *Journal of the Acoustical Society of America* 101:3136.

Rivers, J.A. 1997. Blue whale, *Balaenoptera musculus*, vocalizations from the waters off central California. *Marine Mammal Science* 13(2):186-195.

Ross, D. 1976. *Mechanics of Underwater Noise.* Pergamon, New York.

Saunders, J.C., Y.E. Cohen, and Y.M. Szymko. 1991. The structural and functional consequences of acoustic injury in the cochlea and peripheral auditory system: A five year update. *Journal of the Acoustical Society of America* 90:147-155.

Schlundt, C.E., J.J. Finneran, D.A. Carder, and S.H. Ridgway. 2000. Temporary shift in masked hearing thresholds (MTTS) of bottlenose dolphins, *Tursiops truncatus*, and white whales, *Delphinapterus leucas*, after exposure to intense tones. *Journal of the Acoustical Society of America* (in press).

Simmonds, M.P., and L.F. Lopez-Jurado. 1991. Whales and the military. *Nature* 351:448.

Stafford, K.M., and C.G. Fox. 1996. Occurrence of blue and fin whales calls in the north Pacific as monitored by U.S. Navy SOSUS arrays. *Journal of the Acoustical Society of America* 100:2611. Abstract.

Stafford, K.M., C.G. Fox, and D.S. Clark. 1998. Long-range acoustic detection and localization of blue whale calls in the northeast Pacific Ocean. *Journal of the Acoustical Society of America* 104:3616-3625.

Stafford, K.M., S.L. Nieukirk, and C.G. Fox. 1999. An acoustic link between blue whales in the Eastern Tropical Pacific and the Northeast Pacific. *Marine Mammal Science* 15(4):1258-1268.

Stephens, S.D.G., and H.M. Ballam. 1974. The sono-ocular test. *The Journal of Laryngology and Otology* 88:1049-1059.

Swartz, R.L., and R.J. Hofman. 1991. *Marine Mammal and Habitat Monitoring: Requirements, Principles, Needs, and Approaches.* Report prepared for Marine Mammal Commission. NTIS PB91-215046. National Technical Information Service, Springfield, VA.

Szymanski, M.D., D.E. Bain, and K.R. Henry. 1995. Auditory evoked potentials of killer whale (*Orcinus orca*). Pp. 1-10 in *Sensory Systems of Aquatic Mammals*, R.A. Kastelein, J.A. Thomas, and P.E. Nachtigall (eds.). De Spil Publishers, Woerden, The Netherlands.

Szymanski, M.D., D. Supin, A. Ya, D.E. Bain, and K.R. Henry. 1998. Killer whale (*Orcinus orca*) auditory evoked potentials to rhythmic clicks. *Marine Mammal Science* 14:676-691.

Szymanski, M.D., D.E. Bain, and K. Kiehl. 1999. Killer whale (*Orcinus orca*) hearing: Auditory brainstem response and behavioral audiograms. *Journal of the Acoustical Society of America* 106:1134-1141.

Terhune, J., and S. Turnbull. 1995. Variation in the psychometric functions and hearing thresholds of a harbour seal. Pp. 81-92 in *Sensory Systems of Aquatic Mammals*, R.A. Kastelien, J.A. Thomas, and P.E. Nachtigall (eds.). De Spil Publishers, Woerden, The Netherlands.

Thomas, J.A., S.R. Fisher, L.M. Ferm, and R.S. Hart. 1986. Acoustic detection of cetaceans using a towed array of hydrophones. *Reports of the International Whaling Commission* (Special Issue) 8:139-148.

Thompson, P., L. O'Findley, T. Vidal, and W.C. O'Cummings. 1996. Underwater sounds of blue whales, *Balaenoptera musculus*, in the Gulf of California, Mexico. *Marine Mammal Science* 12:288-293.

Tyack, P. 1981. Interactions between singing Hawaiian humpback whales and conspecifics nearby. *Behavioral Ecology and Sociobiology* 8:105-116.

Tyack, P.L. 1983. Differential response of humpback whales, *Megaptera novaeangliae*, to playback of song or social sounds. *Behavioral Ecology and Sociobiology* 18:251-257.

Tyack, P. 1998. Acoustic communication under the sea. Pp. 163-220 in *Animal Acoustic Communication: Recent Technical Advances*, S.L. Hopp, M.J. Owren, and C.S. Evans (eds.). Springer-Verlag, Heidelberg.

Tyack, P.L., and C.W. Clark. 1998. Quick-Look Report: Playback of Low-Frequency Sound to Gray Whales Migrating Past the Central California Coast. Unpublished report.

Urick, R.J. 1983. *Principles of Underwater Sound.* Third Edition. McGraw-Hill, New York.

Urick, R.J. 1986. *Ambient Noise in the Sea.* Peninsula Publishing, Los Altos, CA.

Ward, W.D. 1997. Effects of high-intensity sound. Pp. 1497-1507 in *Encyclopedia of Acoustics*, M.J. Crocker (ed.). John Wiley & Sons, New York.

Watkins, W.A. 1981. Activities and underwater sounds of fin whales. *Scientific Reports of the Whales Research Institute* 33:83-117.

Watkins, W.A., and W.E. Schevill. 1975. Sperm whales (*Physeter catadon*) react to pingers. *Deep-Sea Research* 22(3):123-129.

Watkins, W.A., K.E. Moore, and P. Tyack. 1985. Sperm whale acoustic behaviors in the southeast Caribbean. *Cetology* 49:1-15.

Watkins, W.A., M.A. Daher, K.M. Fristrup, T.J. Howald, and G. Notarbartolo di Sciara. 1993. Sperm whales tagged with transponders and tracked underwater by sonar. *Marine Mammal Science* 9(1):55-67.

Watkins, W.A., M.A. Daher, G.M. Reppucci, J.E. George, D.L. Martin, N.A. DiMarzio, and D.P. Gannon. 2000. Seasonality and distribution of whale calls in the North Pacific. *Oceanography* 13:62-67.

Yost, W.A., and M.C. Killion. 1997. Hearing thresholds. Pp. 1545-1554 in *Encyclopedia of Acoustics*, M.J. Crocker (ed.). Wiley, New York.

Appendixes

Arthur N. Popper earned his Ph.D. in biology from the City University of New York in 1969. His research interests include vertebrate hearing; structure, function, and evolution of the ear; development of ear and particularly of sensory hair cells; plasticity in the vertebrate auditory system; and innervation of the ear. Dr. Popper has been a professor in the Department of Biology at the University of Maryland, College Park, since 1987.

Harry A. DeFerrari earned his Ph.D. from Catholic University of America in 1966. His research has been in the area of ocean acoustics and sound propagation. Dr. DeFerrari has been a professor at the University of Miami's Rosenstiel School of Marine and Atmospheric Sciences since 1967.

William F. Dolphin earned his Ph.D. in biology from Boston University in 1988. His research interests include auditory physiology and information processing, sensory biophysics, and biosonar. Dr. Dolphin has been a research assistant professor in the Department of Biomedical Engineering and the Department of Biology at Boston University since 1991.

Peggy L. Edds-Walton earned her Ph.D. in zoology from the University of Maryland, College Park, in 1994. Her research interests include vocalizations and behavior of baleen whales and auditory processing in fish. Dr. Edds-Walton is currently a research associate at the Parmly Hearing Institute and a summer scientist at the Marine Biological Laboratory, Woods Hole, Massachusetts.

Gordon M. Greve earned his Ph.D. in geophysics from Stanford University in 1962. He began working for Amoco Production Company (now BP-Amoco) in 1960 and was Manager of Geophysical Research from 1980 to 1986, when he became Manager of Geophysics. Dr. Greve has been a consultant specializing in geophysical methods applied to petroleum exploration since retiring from Amoco in 1994.

Dennis McFadden earned his Ph.D. in sensory psychology from Indiana University in 1967. His research interests include sex and ear differences in hearing and temporary hearing loss induced by drugs and by exposure to intense sounds. Dr. McFadden has been a faculty member in the Department of Psychology at the University of Texas, Austin, since 1967, and is currently an Ashbel Smith Professor.

Peter B. Rhines earned his Ph.D. from Trinity College, Cambridge University, in England in 1967. His research interests include circulation of the oceans; waves, eddies and currents; and climate and transport of natural and artificial trace chemicals in the seas. Dr. Rhines has been a professor of oceanography and atmospheric sciences at the University of Washington since 1984.

Sam H. Ridgway earned his Ph.D. from University College (now Wolfson College), Cambridge University, in 1973. He received a Doctor of Veterinary Medicine Degree (DVM) from Texas A&M University in 1960. His research interests include marine mammal physiology (especially diving and hearing), dolphin neurobiology, and aquatic animal medicine. Dr. Ridgway has been with the U.S. Navy Marine Mammal Program since 1962.

Robert M. Seyfarth earned his Ph.D. from the University of Cambridge in 1976. His research interests include auditory mechanisms and acoustic behavior of nonmarine mammals. Dr. Seyfarth has been a professor in the Department of Psychology at the University of Pennsylvania since 1985.

Sharon L. Smith earned her Ph.D. in zoology from Duke University in 1975. Her research interests include ecology of zooplankton, herbivorous crustaceans, food chain dynamics, and biochemical cycling in productive areas of the ocean. Dr. Smith has been a professor at the University of Miami's Rosenstiel School of Marine and Atmospheric Sciences since 1993, before which she worked at the U.S. Department of Energy's Brookhaven National Laboratory.

Peter L. Tyack earned his Ph.D. in animal behavior from Rockefeller University in 1982. His research interests include cetacean social behavior and vocalizations. Dr. Tyack has been a senior scientist at the Woods Hole Oceanographic Institution since 1999.

B

Summary from NRC (1994)

LIMITATIONS OF CURRENT KNOWLEDGE

Data on the effects of low-frequency sounds on marine mammals are scarce. Although we do have some knowledge about the behavior and reactions of certain marine mammals in response to sound, as well as about the hearing capabilities of a few species, the data are extremely limited and cannot constitute the basis for informed prediction or evaluation of the effects of intense low-frequency sounds on any marine species.

The committee could find almost no quantitative information with which to assess the impact of low-frequency noise on marine mammals. For those few marine mammals on which data are available about their hearing sensitivity, it appears that low-frequency sound, even at very high levels, is barely audible to them. In addition, the range of frequencies by which these animals are affected appears to vary *among,* as well as *within,* the three different orders of Mammalia to which they belong. Certainly data on the hearing sensitivities of several Odontoceti (examples include the white whale, bottlenose dolphin, harbor porpoise, and false killer whale) and Pinnipedia (for example, several seals and the California sea lion) suggest that sounds below about 100 Hz are practically inaudible to these mammals. But even these data are extremely limited and cannot be used to evaluate the effects of intense low-frequency sounds on all species of marine mammals.

There have been some observational or experimental studies and numerous anecdotal reports about the responses of marine mammals to certain sounds. Rather than summarize the existing reviews, the committee decided that its efforts could be more usefully directed to a discussion of the implications of the existing information. The committee noted, for example, that missing in most of these anecdotal accounts is information on the level of the sound exposure experienced by individual animals. Typically, neither the source level nor the received level was measured. Even when the approximate level at the source was known, the received level near the animal was usually not measured, and if it was, there were often uncertainties associated with calculating that level.

This dearth of scientific evidence makes it virtually impossible to predict the effects of low-frequency sound on marine mammals, especially on baleen whales. In the absence of such an impact assessment, the committee finds itself unable to fulfill the second part of its charge, namely, to balance the costs and the benefits of "underwater sound as a research tool" versus "the possibility of harmful effects to marine mammals." Rigorous experimental research on marine mammals and their major prey is required to resolve the issue of how low-frequency sound affects these species. The committee recommends that future experiments be conducted in such a manner that the received level of the sound and the behavior of the animal can be studied together. Such investigations may be logistically complex and may require permits, which are sometimes difficult to obtain.

CHANGES PROPOSED IN REGULATORY STRUCTURE

It is the committee's judgment that the regulatory system governing marine mammal "taking" by research actively discourages and delays the acquisition of scientific knowledge that would benefit conservation of marine mammals, their food sources, and their ecosystems. The committee thus proposes several alternatives for reducing unnecessary regulatory barriers and facilitating valuable research while maintaining all necessary protection for marine mammals.

Although the committee strongly agrees with and supports the objective of marine mammal conservation, it believes that the present regulation of research is unnecessarily cumbersome and restrictive. Not only is research hampered, but the process of training and employing scientists with suitable research skills is impeded by this system. Better and more humane management of marine mammals depends on understanding them better. Well-trained researchers are the ultimate source of our knowledge about marine mammals. The present system, in effect, impedes acquisition of the information and understanding needed to pursue a more effective conservation policy.

The committee considered several possible alternatives for facilitating valuable research while maintaining all necessary protection for marine mammals. One alternative would be to incorporate scientific researchers as "other users" in the regulatory regime recently proposed by the National Marine Fisheries Service (NMFS) of the U.S. Department of Commerce to govern commercial fishing and marine mammal interactions. Another alternative would be to establish a decentralized regulatory regime, possibly patterned after the Institutional Animal Care and Use Committee (IACUC) system currently used to monitor research conducted on nonmarine animals in scientific laboratories.

If the existing system of regulations is maintained, the committee urges that steps be taken to expedite the small incidental take authorization process for all scientific activities involving nonlethal takes, and to further simplify the process for nonlethal takes producing only negligible impact. The committee suggests rewording those provisions to delete references to effects on "small" numbers of marine mammals, provided that the effects are negligible. It would also be beneficial to broaden the definition of research for which scientific permits can be issued to include activities beyond those directly "on or benefiting marine mammals." In order to place regulations on a more rational footing, the population status of each species should determine the number and types of allowable takes, and the same regulations should apply equally to all activities, scientific and otherwise. The committee notes that some of these recommendations would require congressional action to change the Marine Mammal Protection Act and perhaps other laws. However, other recommendations could be implemented under existing laws through changes in regulations.

The committee is by no means recommending a blanket waiver of the requirements for scientific research under the Marine Mammal Protection Act, the Endangered Species Act, and the National Environmental Policy Act—whether on marine mammals or on other topics where experiments might incidentally affect marine mammals. Rather, the committee urges a more logical balance between the regulation of research and other human activities, and a more expeditious permitting process. Appropriate scientific research might identify the sources of human-made noise that actually endanger marine mammals, and may suggest regulation of certain sound sources that are presently unregulated. This research could provide information that would benefit all marine mammals.

Finally, the committee considered the "120-decibel (dB) criterion" that is regarded in some contexts as a noise level above which potentially harmful acoustic effects on marine mammals might occur. In the opinion of the committee, the data from which the 120-dB criterion was derived are being overly extrapolated, largely because of the scarcity of experiments providing quantitative information about the behavior of marine mammals in relation to sound exposure. It is possible that this level is simply the one at which the animals detected the presence of a sound. If this is true, then there is no scientific evidence to indicate that the relatively minor and short-term behavioral reactions observed indicate any significant or long-term effects on the animals. Marine mammals, like other animals, respond to many stimuli, natural and human-made. These reactions are part of their normal behavioral repertoire and are not necessarily indicative of an adverse effect.

One danger of adopting a single number, as with the 120-dB criterion, is in applying it to all species of marine mammals and to all sounds and situations, regardless of the frequency spectrum, regardless of the temporal pattern of the sound, and regardless of differences in the auditory sensitivity of the different groups of marine mammals. There is general agreement that these variables are important in determining whether the 120-dB figure is appropriate in any given situation.

RECOMMENDED RESEARCH

The research that would provide some of the missing information is conceptually straightforward biological science, the proposed experiments should provide much of the needed information, and the cost is not enormous compared with that of other scientific efforts of comparable magnitude.

The committee's aim was to identify general research needs that are crucial to a full evaluation of the effects of intense low-frequency sounds on a variety of marine mammals and their major prey. The committee has identified the following general areas in which more information must be developed:

1. **Research on the behavior of marine mammals in the wild.**
2. **Research on the auditory systems of marine mammals.**
3. **Research on the effects of low-frequency sound on the food chain of marine mammals.**
4. **Development and application of measurement techniques to enhance observation and data gathering.**

The committee recommends that an accelerated program of scientific studies of the acoustic effects on marine mammals and their major prey be undertaken. These studies should be designed to provide information needed to direct policies that will provide long-term protection to the species.

C Relevant U.S. Legislation and Regulations for Marine Mammals

MARINE MAMMAL PROTECTION ACT (16 U.S.C. 31)[1]
(SELECTED PORTIONS)

Sec. 1362. Definitions

For the purposes of this chapter—

(1) The term "depletion'" or "depleted" means any case in which—

(A) the Secretary, after consultation with the Marine Mammal Commission and the Committee of Scientific Advisors on Marine Mammals established under subchapter III of this chapter, determines that a species or population stock is below its optimum sustainable population;

(B) a State, to which authority for the conservation and management of a species or population stock is transferred under section 1379 of this title, determines that such species or stock is below its optimum sustainable population; or

(C) a species or population stock is listed as an endangered species or a threatened species under the Endangered Species Act of 1973 [16 U.S.C. 1531 et seq.].

(2) The terms "conservation'" and "management" mean the collection and application of biological information for the purposes of increasing and maintaining the number of animals within species and populations of marine mammals at their optimum sustainable population. Such terms include the entire scope of activities that constitute a modern scientific resource program, including, but not limited to, research, census, law enforcement, and

---

[1]www4.law.cornell.edu/uscode/16/ch31.html, accessed 9/6/99.

habitat acquisition and improvement. Also included within these terms, when and where appropriate, is the periodic or total protection of species or populations as well as regulated taking.

(6) The term "marine mammal" means any mammal which—

(A) is morphologically adapted to the marine environment including sea otters and members of the orders Sirenia, Pinnipedia, and Cetacea, or

(B) primarily inhabits the marine environment (such as the polar bear); and, for the purposes of this chapter, includes any part of any such marine mammal, including its raw, dressed, or dyed fur or skin.

(7) The term "marine mammal product" means any item of merchandise which consists, or is composed in whole or in part, of any marine mammal.

(8) The term "moratorium" means a complete cessation of the taking of marine mammals and a complete ban on the importation into the United States of marine mammals and marine mammal products, except as provided in this chapter.

(9) The term "optimum sustainable population" means, with respect to any population stock, the number of animals which will result in the maximum productivity of the population or the species, keeping in mind the carrying capacity of the habitat and the health of the ecosystem of which they form a constituent element.

(10) The term "person" includes (A) any private person or entity, and (B) any officer, employee, agent, department, or instrumentality of the Federal Government, of any State or political subdivision thereof, or of any foreign government.

(11) The term "population stock" or "stock" means a group of marine mammals of the same species or smaller taxa in a common spatial arrangement, that interbreed when mature.

(A) Except as provided in subparagraph (B), the term "Secretary" means—

(i) the Secretary of the department in which the National Oceanic and Atmospheric Administration is operating, as to all responsibility, authority, funding, and duties under this chapter with respect to members of the order Cetacea and members, other than walruses, of the order Pinnipedia, and

(ii) the Secretary of the Interior as to all responsibility, authority, funding, and duties under this chapter with respect to all other marine mammals covered by this chapter.

(B) in section 1387 of this title and subchapter V of this chapter the term "Secretary" means the Secretary of Commerce.

(13) The term "take" means to harass, hunt, capture or kill, or attempt to harass, hunt, capture, or kill any marine mammal.

(14) The term "United States" includes the several States, the District of

Columbia, the Commonwealth of Puerto Rico, the Virgin Islands of the United States, American Samoa, Guam, and Northern Mariana Islands.

(15) The term "waters under the jurisdiction of the United States" means—

(A) the territorial sea of the United States, and

(B) the waters included within a zone, contiguous to the territorial sea of the United States, of which the inner boundary is a line coterminous with the seaward boundary of each coastal State, and the outer boundary is a line drawn in such a manner that each point on it is 200 nautical miles from the baseline from which the territorial sea is measured.

(16) The term "fishery" means—

(A) one or more stocks of fish which can be treated as a unit for purposes of conservation and management and which are identified on the basis of geographical, scientific, technical, recreational, and economic characteristics; and

(B) any fishing for such stocks.

(18)

(A) The term "harassment" means any act of pursuit, torment, or annoyance which

(i) has the potential to injure a marine mammal or marine mammal stock in the wild; or

(ii) has the potential to disturb a marine mammal or marine mammal stock in the wild by causing disruption of behavioral patterns, including, but not limited to, migration, breathing, nursing, breeding, feeding, or sheltering.

(B) The term "Level A harassment" means harassment described in subparagraph (A)(i).

(C) The term "Level B harassment" means harassment described in subparagraph (A)(ii).

(19) The term "strategic stock" means a marine mammal stock—

(A) for which the level of direct human-caused mortality exceeds the potential biological removal level;

(B) which, based on the best available scientific information, is declining and is likely to be listed as a threatened species under the Endangered Species Act of 1973 [16 U.S.C. 1531 et seq.] within the foreseeable future; or

(C) which is listed as a threatened species or endangered species under the Endangered Species Act of 1973 (16 U.S.C. 1531 et seq.), or is designated as depleted under this chapter.

(20) The term "potential biological removal level" means the maximum number of animals, not including natural mortalities, that may be removed from a marine mammal stock while allowing that stock to reach or maintain its optimum sustainable population. The potential biological removal level is the product of the following factors:

(A) The minimum population estimate of the stock.

(B) One-half the maximum theoretical or estimated net productivity rate of the stock at a small population size.

(C) A recovery factor of between 0.1 and 1.0.

Sec. 1371. Moratorium on taking and importing marine mammals and marine mammal products

(a) Imposition; exceptions. There shall be a moratorium on the taking and importation of marine mammals and marine mammal products, commencing on the effective date of this chapter, during which time no permit may be issued for the taking of any marine mammal and no marine mammal or marine mammal product may be imported into the United States except in the following cases:

(1) Consistent with the provisions of section 1374 of this title, permits may be issued by the Secretary for taking, and importation for purposes of scientific research, public display, photography for educational or commercial purposes, or enhancing the survival or recovery of a species or stock, or for importation of polar bear parts (other than internal organs) taken in sport hunts in Canada. Such permits, except permits issued under section 1374(c)(5) of this title, may be issued if the taking or importation proposed to be made is first reviewed by the Marine Mammal Commission and the Committee of Scientific Advisors on Marine Mammals established under subchapter III of this chapter. The Commission and Committee shall recommend any proposed taking or importation, other than importation under section 1374(c)(5) of this title, which is consistent with the purposes and policies of section 1361 of this title. If the Secretary issues such a permit for importation, the Secretary shall issue to the importer concerned a certificate to that effect in such form as the Secretary of the Treasury prescribes, and such importation may be made upon presentation of the certificate to the customs officer concerned.

(3)

(A) The Secretary, on the basis of the best scientific evidence available and in consultation with the Marine Mammal Commission, is authorized and directed, from time to time, having due regard to the distribution, abundance, breeding habits, and times and lines of migratory movements of such marine mammals, to determine when, to what extent, if at all, and by what means, it is compatible with this chapter to waive the requirements of this section so as to allow taking, or importing of any marine mammal, or any marine mammal product, and to adopt suitable regulations, issue permits, and make determinations in accordance with sections 1372, 1373, 1374, and 1381 of this title permitting and governing such taking and importing, in accordance with such determinations: Provided, however, that the Secretary, in making such determinations must be assured that the taking of such marine mammal is in accord with sound principles of resource protection and conservation as provided in the purposes and policies of this chapter: Provided, further, however, that no marine mammal or no marine mammal product may be imported into the United States unless the Secretary certifies that the program for taking marine mammals in the country of origin is consistent with the provisions and policies of this chapter. Products of nations not so certified may not be imported into the United States for any purpose, including processing for exportation. (B) Except for scientific research purposes, photography for educational or commercial purposes, or enhancing the survival or recovery of a species or stock as provided for in paragraph (1) of this subsection, or as provided for under paragraph (5) of this subsection, during the moratorium no permit may be issued for the taking of any marine mammal which has been designated by the Secretary as depleted, and no importation may be made of any such mammal.

(4)

(A) Except as provided in subparagraphs (B) and (C), the provisions of this chapter shall not apply to the use of measures—

(i) by the owner of fishing gear or catch, or an employee or agent of such owner, to deter a marine mammal from damaging the gear or catch;

(ii) by the owner of other private property, or an agent, bailee, or employee of such owner, to deter a marine mammal from damaging private property;

(iii) by any person, to deter a marine mammal from endangering personal safety; or

(iv) by a government employee, to deter a marine mammal from damaging public property, so long as such measures do not result in the death or serious injury of a marine mammal.

(5)

(A) Upon request therefor by citizens of the United States who engage in a specified activity (other than commercial fishing) within a specified geographical region, the Secretary shall allow, during periods of not more than five consecutive years each, the incidental, but not intentional, taking by citizens while engaging in that activity within that region of small numbers of marine mammals of a species or population stock if the Secretary, after notice (in the Federal Register and in newspapers of general circulation, and through appropriate electronic media, in the coastal areas that may be affected by such activity) and opportunity for public comment—

(i) finds that the total of such taking during each five-year (or less) period concerned will have a negligible impact on such species or stock and will not have an unmitigable adverse impact on the availability of such species or stock for taking for subsistence uses pursuant to subsection (b) of this section or section 1379(f) of this title or, in the case of a cooperative agreement under both this chapter and the Whaling Convention Act of 1949 (16 U.S.C. 916 et seq.), pursuant to section 1382 of this title; and

(ii) prescribes regulations setting forth—

(I) permissible methods of taking pursuant to such activity, and other means of effecting the least practicable adverse impact on such species or stock and its habitat, paying particular attention to rookeries, mating grounds, and areas of similar significance, and on the availability of such species or stock for subsistence uses; and

(II) requirements pertaining to the monitoring and reporting of such taking.

(B) The Secretary shall withdraw, or suspend for a time certain (either on an individual or class basis, as appropriate) the permission to take marine mammals under subparagraph (A) pursuant to a specified activity within a specified geographical region if the Secretary finds, after notice and opportunity for public comment (as required under subparagraph (A) unless subparagraph (C)(I) applies), that—

(i) the regulations prescribed under subparagraph (A) regarding methods of taking, monitoring, or reporting are not being substantially complied with by a person engaging in such activity; or

(ii) the taking allowed under subparagraph (A) pursuant to one or more activities within one or more regions is having, or may have, more than a negligible impact on the species or stock concerned.

(D)

(i) Upon request therefor by citizens of the United States who engage in a specified activity (other than commercial fishing) within a specific geographic region, the Secretary shall authorize, for periods of not more than 1 year, subject to such conditions as the Secretary may specify, the incidental, but not intentional, taking by harassment of small numbers of marine mammals of a species or population stock by such citizens while engaging in that activity within that region if the Secretary finds that such harassment during each period concerned—

(I) will have a negligible impact on such species or stock, and

(II) will not have an unmitigable adverse impact on the availability of such species or stock for taking for subsistence uses pursuant to subsection (b) of this section, or section 1379(f) of this title or pursuant to a cooperative agreement under section 1388 of this title.

(ii) The authorization for such activity shall prescribe, where applicable—

(I) permissible methods of taking by harassment pursuant to such activity, and other means of effecting the least practicable impact on such species or stock and its habitat, paying particular attention to rookeries, mating grounds, and areas of similar significance, and on the availability of such species or stock for taking for subsistence uses pursuant to subsection (b) of this section or section 1379(f) of this title or pursuant to a cooperative agreement under section 1388 of this title,

(II) the measures that the Secretary determines are necessary to ensure no unmitigable adverse impact on the availability of the species or stock for taking for subsistence uses pursuant to subsection (b) of this section or section 1379(f) of this title or pursuant to a cooperative agreement under section 1388 of this title, and

(III) requirements pertaining to the monitoring and reporting of such taking by harassment, including requirements for the independent peer review of proposed monitoring plans or other research proposals where the proposed activity may affect the availability of a species or stock for taking for subsistence uses pursuant to subsection (b) of this section or section 1379(f) of this title or pursuant to a cooperative agreement under section 1388 of this title.

(iii) The Secretary shall publish a proposed authorization not later than 45 days after receiving an application under this subparagraph and request public comment through notice in the Federal Register, newspapers of general circulation, and appropriate electronic media and to all locally affected communities for a period of 30 days after publication. Not later than 45 days after the close of the public comment period, if the Secretary makes the findings set forth in clause (i), the Secretary shall issue an authorization with appropriate conditions to meet the requirements of clause (ii).

(iv) The Secretary shall modify, suspend, or revoke an authori-

zation if the Secretary finds that the provisions of clauses (i)
or (ii) are not being met.

(v) A person conducting an activity for which an authorization
has been granted under this subparagraph shall not be subject
to the penalties of this chapter for taking by harassment that
occurs in compliance with such authorization.

Sec. 1374. Permits

(a) Issuance—The Secretary may issue permits which authorize the taking or
importation of any marine mammal. Permits for the incidental taking of
marine mammals in the course of commercial fishing operations may only be
issued as specifically provided for in sections 1371(a)(5) or 1416 of this title,
or subsection (h) of this section.

(b) Requisite provisions—Any permit issued under this section shall—
    (1) be consistent with any applicable regulation established by the Sec-
retary under section 1373 of this title, and
    (2) specify—
        (A) the number and kind of animals which are authorized to be
taken or imported,
        (B) the location and manner (which manner must be determined by
the Secretary to be humane) in which they may be taken, or from
which they may be imported,
        (C) the period during which the permit is valid, and
        (D) any other terms or conditions which the Secretary deems appro-
priate. In any case in which an application for a permit cites as a
reason for the proposed taking the overpopulation of a particular
species or population stock, the Secretary shall first consider
whether or not it would be more desirable to transplant a number of
animals (but not to exceed the number requested for taking in the
application) of that species or stock to a location not then inhabited
by such species or stock but previously inhabited by such species or
stock.

(c) Importation for scientific research, public display, or enhancing survival
or recovery of species or stock—
    (1) Any permit issued by the Secretary which authorizes the taking or
importation of a marine mammal for purposes of scientific research,
public display, or enhancing the survival or recovery of a species or
stock shall specify, in addition to the conditions required by subsection
(b) of this section, the methods of capture, supervision, care, and transportation which must be observed pursuant to such taking or importa-

tion. Any person authorized to take or import a marine mammal for purposes of scientific research, public display, or enhancing the survival or recovery of a species or stock shall furnish to the Secretary a report on all activities carried out by him pursuant to that authority.

(2)

(B) A permit under this paragraph shall grant to the person to which it is issued the right, without obtaining any additional permit or authorization under this chapter, to—

(i) take, import, purchase, offer to purchase, possess, or transport the marine mammal that is the subject of the permit; and

(ii) sell, export, or otherwise transfer possession of the marine mammal, or offer to sell, export, or otherwise transfer possession of the marine mammal—

(II) for the purpose of scientific research, to a person that meets the requirements of paragraph (3); or

(III) for the purpose of enhancing the survival or recovery of a species or stock, to a person that meets the requirements of paragraph (4).

(C) A person to which a marine mammal is sold or exported or to which possession of a marine mammal is otherwise transferred under the authority of subparagraph (B) shall have the rights and responsibilities described in subparagraph (B) with respect to the marine mammal without obtaining any additional permit or authorization under this chapter. Such responsibilities shall be limited to—

(ii) for the purpose of scientific research, the responsibility to meet the requirements of paragraph (3), and

(iii) for the purpose of enhancing the survival or recovery of a species or stock, the responsibility to meet the requirements of paragraph (4).

(E) No marine mammal held pursuant to a permit issued under subparagraph (A), or by a person exercising rights under subparagraph (C), may be sold, purchased, exported, or transported unless the Secretary is notified of such action no later than 15 days before such action, and such action is for purposes of public display, scientific research, or enhancing the survival or recovery of a species or stock. The Secretary may only require the notification to include the information required for the inventory established under paragraph (10).

(3)

(A) The Secretary may issue a permit under this paragraph for scientific research purposes to an applicant which submits with its permit application information indicating that the taking is required to further a bona fide scientific purpose. The Secretary may issue a permit under this paragraph before the end of the public review and comment period required under subsection (d)(2) of this section if delaying issuance of the permit could result in injury to a species, stock, or individual, or in loss of unique research opportunities.

(B) No permit issued for purposes of scientific research shall authorize the lethal taking of a marine mammal unless the applicant demonstrates that a nonlethal method of conducting the research is not feasible. The Secretary shall not issue a permit for research which involves the lethal taking of a marine mammal from a species or stock that is depleted, unless the Secretary determines that the results of such research will directly benefit that species or stock, or that such research fulfills a critically important research need.

(C) Not later than 120 days after April 30, 1994, the Secretary shall issue a general authorization and implementing regulations allowing bona fide scientific research that may result only in taking by Level B harassment of a marine mammal. Such authorization shall apply to persons which submit, by 60 days before commencement of such research, a letter of intent via certified mail to the Secretary containing the following:

(i) The species or stocks of marine mammals which may be harassed.

(ii) The geographic location of the research.

(iii) The period of time over which the research will be conducted.

(iv) The purpose of the research, including a description of how the definition of bona fide research as established under this chapter would apply.

(v) Methods to be used to conduct the research. Not later than 30 days after receipt of a letter of intent to conduct scientific research under the general authorization, the Secretary shall issue a letter to the applicant confirming that the general authorization applies, or, if the proposed research is likely to result in the taking (including Level A harassment) of a marine mammal, shall notify the applicant that subparagraph (A) applies.

(d) Application procedures; notice; hearing; review—

(1) The Secretary shall prescribe such procedures as are necessary to carry out this section, including the form and manner in which application for permits may be made.

(2) The Secretary shall publish notice in the Federal Register of each application made for a permit under this section. Such notice shall invite the submission from interested parties, within thirty days after the date

of the notice, of written data or views, with respect to the taking or importation proposed in such application.

(3) The applicant for any permit under this section must demonstrate to the Secretary that the taking or importation of any marine mammal under such permit will be consistent with the purposes of this chapter and the applicable regulations established under section 1373 of this title.

(4) If within thirty days after the date of publication of notice pursuant to paragraph (2) of this subsection with respect to any application for a permit any interested party or parties request a hearing in connection therewith, the Secretary may, within sixty days following such date of publication, afford to such party or parties an opportunity for such a hearing.

(5) As soon as practicable (but not later than thirty days) after the close of the hearing or, if no hearing is held, after the last day on which data, or views, may be submitted pursuant to paragraph (2) of this subsection, the Secretary shall

(A) issue a permit containing such terms and conditions as he deems appropriate, or

(B) shall deny issuance of a permit. Notice of the decision of the Secretary to issue or to deny any permit under this paragraph must be published in the Federal Register within ten days after the date of issuance or denial.

(6) Any applicant for a permit, or any party opposed to such permit, may obtain judicial review of the terms and conditions of any permit issued by the Secretary under this section or of his refusal to issue such a permit. Such review, which shall be pursuant to chapter 7 of title 5, may be initiated by filing a petition for review in the United States district court for the district wherein the applicant for a permit resides, or has his principal place of business, or in the United States District Court for the District of Columbia, within sixty days after the date on which such permit is issued or denied.

(g) Fees—The Secretary shall establish and charge a reasonable fee for permits issued under this section.

50 CFR PART 216—REGULATIONS GOVERNING THE TAKING AND IMPORTING OF MARINE MAMMALS (SELECTED PORTIONS)

Subpart A – Introduction

§ 216.3 Definitions.

In addition to definitions contained in the MMPA, and unless the context otherwise requires, in this part 216:

Acts means, collectively, the Marine Mammal Protection Act of 1972, as amended, 16 U.S.C. 1361 et seq., the Endangered Species Act of 1973, as amended, 16 U.S.C. 1531 et seq., and the Fur Seal Act of 1966, as amended, 16 U.S.C. 1151 et seq.

Bona fide scientific research:

(1) Means scientific research on marine mammals conducted by qualified personnel, the results of which:

(i) Likely would be accepted for publication in a refereed scientific journal;

(ii) Are likely to contribute to the basic knowledge of marine mammal biology or ecology (Note: This includes, for example, marine mammal parts in a properly curated, professionally accredited scientific collection); or

(iii) Are likely to identify, evaluate, or resolve conservation problems.

(2) Research that is not on marine mammals, but that may incidentally take marine mammals, is not included in this definition (see sections 101(a)(3)(A), 101(a)(5)(A), and 101(a)(5)(D) of the MMPA, and sections 7(b)(4) and 10(a)(1)(B) of the ESA).

Intrusive research means a procedure conducted for bona fide scientific research involving: A break in or cutting of the skin or equivalent, insertion of an instrument or material into an orifice, introduction of a substance or object into the animal's immediate environment that is likely either to be ingested or to contact and directly affect animal tissues (i.e., chemical substances), or a stimulus directed at animals that may involve a risk to health or welfare or that may have an impact on normal function or behavior (i.e., audio broadcasts directed at animals that may affect behavior). For captive animals, this definition does not include:

(1) A procedure conducted by the professional staff of the holding facility or an attending veterinarian for purposes of animal husbandry, care, maintenance, or treatment, or a routine medical procedure that, in the reasonable judgment of the attending veterinarian, would not constitute a risk to the health or welfare of the captive animal; or (2) A procedure

involving either the introduction of a substance or object (i.e., as described in this definition) or a stimulus directed at animals that, in the reasonable judgment of the attending veterinarian, would not involve a risk to the health or welfare of the captive animal.

Level A Harassment means any act of pursuit, torment, or annoyance which has the potential to injure a marine mammal or marine mammal stock in the wild.

Level B Harassment means any act of pursuit, torment, or annoyance which has the potential to disturb a marine mammal or marine mammal stock in the wild by causing disruption of behavioral patterns, including, but not limited to, migration, breathing, nursing, breeding, feeding, or sheltering but which does not have the potential to injure a marine mammal or marine mammal stock in the wild.

Stranded or stranded marine mammal means a marine mammal specimen under the jurisdiction of the Secretary:

(1) If the specimen is dead, and is on a beach or shore, or is in the water within the Exclusive Economic Zone of the United States; or

(2) If the specimen is alive, and is on a beach or shore and is unable to return to the water, or is in the water within the Exclusive Economic Zone of the United States where the water is so shallow that the specimen is unable to return to its natural habitat under its own power.

Take means to harass, hunt, capture, collect, or kill, or attempt to harass, hunt, capture, collect, or kill any marine mammal. This includes, without limitation, any of the following: The collection of dead animals, or parts thereof; the restraint or detention of a marine mammal, no matter how temporary; tagging a marine mammal; the negligent or intentional operation of an aircraft or vessel, or the doing of any other negligent or intentional act which results in disturbing or molesting a marine mammal; and feeding or attempting to feed a marine mammal in the wild.

Subpart B - Prohibitions

§ 216.11 Prohibited taking.

Except as otherwise provided in sub-parts C, D, and I of this part 216 or in part 228 or 229, it is unlawful for:

(a) Any person, vessel, or conveyance subject to the jurisdiction of the United States to take any marine mammal on the high seas, or

(b) Any person, vessel, or conveyance to take any marine mammal in waters or on lands under the jurisdiction of the United States, or

(c) Any person subject to the jurisdiction of the United States to take any marine mammal during the moratorium.

§ 216.16 Prohibitions under the General Authorization for Level B harassment for scientific research.

It shall be unlawful for any person to:

(a) Provide false information in a letter of intent submitted pursuant to § 216.45(b);

(b) Violate any term or condition imposed pursuant to § 216.45(d).

Subpart D—Special Exceptions

§ 216.31 Definitions.

For the purpose of this subpart, the definitions set forth in 50 CFR part 217 shall apply to all threatened and endangered marine mammals, unless a more restrictive definition exists under the MMPA or part 216.

§ 216.32 Scope.

The regulations of this subpart apply to:

(a) All marine mammals and marine mammal parts taken or born in captivity after December 20, 1972; and

(b) All marine mammals and marine mammal parts that are listed as threatened or endangered under the ESA.

§ 216.33 Permit application submission, review, and decision procedures.

(a) Application submission. Persons seeking a special exemption permit under this subpart must submit an application to the Office Director. The application must be signed by the applicant, and provide in a properly formatted manner all information necessary to process the application. Written instructions addressing information requirements and formatting may be obtained from the Office Director upon request.

(c) Initial review.

(1) NMFS will notify the applicant of receipt of the application.

(2) During the initial review, the Office Director will determine:

(i) Whether the application is complete.

(ii) Whether the proposed activity is for purposes authorized under this sub-part.

(iii) If the proposed activity is for enhancement purposes, whether the species or stock identified in the application is in need of enhancement for its survival or recovery and whether the proposed activity will likely succeed in its objectives.

(iv) Whether the activities proposed are to be conducted consistent with the permit restrictions and permit specific conditions as described in § 216.35 and § 216.36(a).

(v) Whether sufficient information is included regarding the environmental impact of the proposed activity to enable the Office Director:

(A) To make an initial determination under the National Environmental Policy Act (NEPA) as to whether the proposed activity is categorically excluded from preparation of further environmental documentation, or whether the preparation of an environmental assessment (EA) or environmental impact statement (EIS) is appropriate or necessary; and

(B) To prepare an EA or EIS if an initial determination is made by the Office Director that the activity proposed is not categorically excluded from such requirements.

(3) The Office Director may consult with the Marine Mammal Commission (Commission) and its Committee of Scientific Advisors on Marine Mammals (Committee) in making these initial, and any subsequent, determinations.

(4) Incomplete applications will be returned with explanation. If the applicant fails to resubmit a complete application or correct the identified deficiencies within 60 days, the application will be deemed withdrawn. Applications that propose activities inconsistent with this subpart will be returned with explanation, and will not be considered further.

(d) Notice of receipt and application review.

(1) Upon receipt of a valid, complete application, and the preparation of any NEPA documentation that has been determined initially to be required, the Office Director will publish a notice of receipt in the FEDERAL REGISTER. The notice will:

(i) Summarize the application, including:

(A) The purpose of the request;

(B) The species and number of marine mammals;

(C) The type and manner of special exception activity proposed;

(D) The location(s) in which the marine mammals will be taken, from which they will be imported, or to which they will be exported; and

(E) The requested period of the permit.

(ii) List where the application is available for review.

(iii) Invite interested parties to submit written comments concerning the application within 30 days of the date of the notice.

(iv) Include a NEPA statement that an initial determination has been made that the activity proposed is categorically excluded from the requirement to prepare an EA or EIS, that an EA was prepared resulting in a finding of no significant impact, or that a final EIS has been prepared and is available for review.

(2) The Office Director will forward a copy of the complete application to the Commission for comment. If no comments are received within 45 days (or such longer time as the Office Director may establish) the Office Director will consider the Commission to have no objection to issuing a permit.

(3) The Office Director may consult with any other person, institution, or agency concerning the application.

(4) Within 30 days of publication of the notice of receipt in the FEDERAL REGISTER, any interested party may submit written comments or may request a public hearing on the application.

(5) If the Office Director deems it advisable, the Office Director may hold a public hearing within 60 days of publication of the notice of receipt in the FEDERAL REGISTER. Notice of the date, time, and place of the public hearing will be published in the FEDERAL REGISTER not less than 15 days in advance of the public hearing. Any interested person may appear in person or through representatives and may submit any relevant material, data, views, or comments. A summary record of the hearing will be kept.

(6) The Office Director may extend the period during which any interested party may submit written comments. Notice of the extension must be published in the FEDERAL REGISTER within 60 days of publication of the notice of receipt in the FEDERAL REGISTER.

(7) If, after publishing a notice of receipt, the Office Director determines on the basis of new information that an EA or EIS must be prepared, the Office Director must deny the permit unless an EA is prepared with a finding of no significant impact. If a permit is denied under these circumstances the application may be resubmitted with information sufficient to prepare an EA or EIS, and will be processed as a new application.

(e) Issuance or denial procedures.

(1) Within 30 days of the close of the public hearing or, if no public hearing is held, within 30 days of the close of the public comment period, the Office Director will issue or deny a special exception permit.

(2) The decision to issue or deny a permit will be based upon:

(i) All relevant issuance criteria set forth at § 216.34;

(ii) All purpose-specific issuance criteria as appropriate set forth at § 216.41, § 216.42, and § 216.43;

(iii) All comments received or views solicited on the permit application; and

(iv) Any other information or data that the Office Director deems relevant.

(3) If the permit is issued, upon receipt, the holder must date and sign the permit, and return a copy of the original to the Office Director. The permit shall be effective upon the permit holder's signing of the permit. In signing the permit, the holder:

(i) Agrees to abide by all terms and conditions set forth in the permit, and all restrictions and relevant regulations under this subpart; and

(ii) Acknowledges that the authority to conduct certain activities specified in the permit is conditional and subject to authorization by the Office Director.

(4) Notice of the decision of the Office Director shall be published in the FEDERAL REGISTER within 10 days after the date of permit issuance or denial and shall indicate where copies of the permit, if issued, may be reviewed or obtained. If the permit issued involves marine mammals listed as endangered or threatened under the ESA, the notice shall include a finding by the Office Director that the permit:

(i) Was applied for in good faith;

(ii) If exercised, will not operate to the disadvantage of such endangered or threatened species; and

(iii) Is consistent with the purposes and policy set forth in section 2 of the ESA.

(5) If the permit is denied, the Office Director shall provide the applicant with an explanation for the denial.

(6) Under the MMPA, the Office Director may issue a permit for scientific research before the end of the public comment period if delaying issuance could result in injury to a species, stock, or individual, or in loss of unique research opportunities. The Office Director also may waive the 30-day comment period required under the ESA in an emergency situation where the health or life of an endangered or threatened marine mammal is threatened and no reasonable alternative is available. If a permit is issued under these circumstances, notice of such issuance before the end of the comment period shall be published in the FEDERAL REGISTER within 10 days of issuance.

(7) The applicant or any party opposed to a permit may seek judicial review of the terms and conditions of such permit or of a decision to deny such permit. Review may be obtained by filing a petition for review with the appropriate U.S. District Court as provided for by law.

§ 216.34 Issuance criteria.

(a) For the Office Director to issue any permit under this subpart, the applicant must demonstrate that:

(1) The proposed activity is humane and does not present any unnecessary risks to the health and welfare of marine mammals;

(2) The proposed activity is consistent with all restrictions set forth at § 216.35 and any purpose-specific restrictions as appropriate set forth at § 216.41, § 216.42, and § 216.43;

(3) The proposed activity, if it involves endangered or threatened marine mammals, will be conducted consistent with the purposes and policies set forth in section 2 of the ESA;

(4) The proposed activity by itself or in combination with other activities, will not likely have a significant adverse impact on the species or stock;

(5) Whether the applicant's expertise, facilities, and resources are adequate to accomplish successfully the objectives and activities stated in the application;

(6) If a live animal will be held captive or transported, the applicant's qualifications, facilities, and resources are adequate for the proper care and maintenance of the marine mammal; and

(7) Any requested import or export will not likely result in the taking of marine mammals or marine mammal parts beyond those authorized by the permit.

(b) The opinions or views of scientists or other persons or organizations knowledgeable of the marine mammals that are the subject of the application or of other matters germane to the application will be considered.

§ 216.35 Permit restrictions.

The following restrictions shall apply to all permits issued under this sub-part:

(a) The taking, importation, export, or other permitted activity involving marine mammals and marine mammal parts shall comply with the regulations of this subpart.

(b) The maximum period of any special exception permit issued, or any major amendment granted, is five years from the effective date of the permit or major amendment. In accordance with the provisions of § 216.39, the period of a permit may be extended beyond that established in the original permit.

(c) Except as provided for in § 216.41(c)(1)(v), marine mammals or marine mammal parts imported under the authority of a permit must be taken or imported in a humane manner, and in compliance with the Acts and any applicable foreign law. Importation of marine mammals and marine mammal parts is subject to the provisions of 50 CFR part 14.

(d) The permit holder shall not take from the wild any marine mammal which at the time of taking is either unweaned or less than eight months old, or is a part of a mother-calf/pup pair, unless such take is specifically authorized in the conditions of the special exception permit. Additionally, the permit holder shall not import any marine mammal that is pregnant or lactating at the time of taking or import, or is unweaned or less than eight months old unless such import is specifically authorized in the conditions of the special exception permit.

(e) Captive marine mammals shall not be released into the wild unless specifically authorized by the Office Director under a scientific research or enhancement permit.

(f) The permit holder is responsible for all activities of any individual who is operating under the authority of the permit;

(g) Individuals conducting activities authorized under the permit must possess qualifications commensurate with their duties and responsibilities, or must be under the direct supervision of a person with such qualifications;

(h) Persons who require state or Federal licenses to conduct activities authorized under the permit must be duly licensed when undertaking such activities;

(i) Special exception permits are not transferable or assignable to any other person, and a permit holder may not require any direct or indirect compensation from another person in return for requesting authorization for such person to conduct the taking, import, or export activities authorized under the subject permit;

(j) The permit holder or designated agent shall possess a copy of the permit when engaged in a permitted activity, when the marine mammal is in transit incidental to such activity, and whenever marine mammals or marine mammal parts are in the possession of the permit holder or agent. A copy of the permit shall be affixed to any container, package, enclosure, or other means of containment, in which the marine mammals or marine mammal parts are placed for purposes of transit, supervision, or care. For marine mammals held captive and marine mammal parts in storage, a copy of the permit shall be kept on file in the holding or storage facility.

§ 216.36 Permit conditions.

(a) Specific conditions.

(1) Permits issued under this subpart shall contain specific terms and conditions deemed appropriate by the Office Director, including, but not limited to:

(i) The number and species of marine mammals that are authorized to be taken, imported, exported, or otherwise affected;

(ii) The manner in which marine mammals may be taken according to type of take;

(iii) The location(s) in which the marine mammals may be taken, from which they may be imported, or to which they may be exported, as applicable, and, for endangered or threatened marine mammal species to be imported or exported, the port of entry or export;

(iv) The period during which the permit is valid.

(b) Other conditions. In addition to the specific conditions imposed pursuant to paragraph (a) of this section, the Office Director shall specify any other permit conditions deemed appropriate.

§ 216.41 Permits for scientific research and enhancement.

In addition to the requirements under §§ 216.33 through 216.38, permits for scientific research and enhancement are governed by the following requirements:

(a) Applicant.

(1) For each application submitted under this section, the applicant shall be the principal investigator responsible for the overall research or enhancement activity. If the research or enhancement activity will involve a periodic change in the principal investigator or is otherwise controlled by and dependent upon another entity, the applicant may be the institution, governmental entity, or corporation responsible for supervision of the principal investigator.

(2) For any scientific research involving captive maintenance, the application must include supporting documentation from the person responsible for the facility or other temporary enclosure.

(b) Issuance Criteria. For the Office Director to issue any scientific research or enhancement permit, the applicant must demonstrate that:

(1) The proposed activity furthers a bona fide scientific or enhancement purpose;

(2) If the lethal taking of marine mammals is proposed:

(i) Non-lethal methods for conducting the research are not feasible; and

(ii) For depleted, endangered, or threatened species, the results will directly benefit that species or stock, or will fulfill a critically important research need.

(3) Any permanent removal of a marine mammal from the wild is consistent with any applicable quota established by the Office Director.

(4) The proposed research will not likely have significant adverse effects on any other component of the marine ecosystem of which the affected species or stock is a part.

(5) For species or stocks designated or proposed to be designated as depleted, or listed or proposed to be listed as endangered or threatened:

(i) The proposed research cannot be accomplished using a species or stock that is not designated or proposed to be designated as depleted, or listed or proposed to be listed as threatened or endangered;

(ii) The proposed research, by itself or in combination with other activities will not likely have a long-term direct or indirect adverse impact on the species or stock;

(iii) The proposed research will either:

(A) Contribute to fulfilling a research need or objective identified in a species recovery or conservation plan, or if there is no conservation or recovery plan in place, a research need or objective identified by the Office Director in stock assessments established under section 117 of the MMPA;

(B) Contribute significantly to understanding the basic biology or ecology of the species or stock, or to identifying, evaluating, or resolving conservation problems for the species or stock; or

(C) Contribute significantly to fulfilling a critically important research need.

(c) Restrictions.

(1) The following restrictions apply to all scientific research permits issued under this sub-part:

(i) Research activities must be conducted in the manner authorized in the permit.

(ii) Research results shall be published or otherwise made available to the scientific community in a reasonable period of time.

(iii) Research activities must be conducted under the direct supervision of the principal investigator or a co-investigator identified in the permit.

(iv) Personnel involved in research activities shall be reasonable in number and limited to:

(A) Individuals who perform a function directly supportive of and necessary to the permitted research activity; and

(B) Support personnel included for the purpose of training or as backup personnel for persons described in paragraph (c)(1)(iv)(A).

(v) Any marine mammal part imported under the authority of a scientific research permit must not have been obtained as the result of a lethal taking that would be inconsistent with the Acts, unless authorized by the Office Director.

(vi) Marine mammals held under a permit for scientific research shall not be placed on public display, included in an interactive program or activity, or trained for performance unless such activities:

(A) Are necessary to address scientific research objectives and have been specifically authorized by the Office Director under the scientific research permit; and

(B) Are conducted incidental to and do not in any way interfere with the permitted scientific research; and

(C) Are conducted in a manner consistent with provisions applicable to public display, unless exceptions are specifically authorized by the Office Director.

(vii) Any activity conducted incidental to the authorized scientific research activity must not involve any taking of marine mammals beyond what is necessary to conduct the research (i.e., educational and commercial photography).

§ 216.44 Applicability/transition.

(a) General. The regulations of this subpart are applicable to all persons, including persons holding permits or other authorizing documents issued before June 10, 1996, by NMFS for the take, import, export, or conduct of any otherwise prohibited activity involving a marine mammal or marine mammal part for special exception purposes.

(b) Scientific research. Any intrusive research as defined in § 216.3, initiated after June 10, 1996, must be authorized under a scientific research permit. Intrusive research authorized by the Office Director to be conducted on captive marine mammals held for public display purposes prior to June 10, 1996, must be authorized under a scientific research permit one year after June 10, 1996.

§ 216.45 General Authorization for Level B harassment for scientific research.

(a) General Authorization.

(1) Persons are authorized under section 104(c)(3)(C) of the MMPA to take marine mammals in the wild by Level B harassment, as defined in § 216.3, for purposes of bona fide scientific research Provided, That:

(i) They submit a letter of intent in accordance with the requirements of paragraph (b) of this section, receive confirmation that the General Authorization applies in accordance with paragraph (c) of this section, and comply with the terms and conditions of paragraph (d) of this section; or

(ii) If such marine mammals are listed as endangered or threatened under the ESA, they have been issued a permit under Section 10(a)(1)(A) of the ESA and implementing regulations at 50 CFR parts 217–227, particularly at § 222.23 through § 222.28, to take marine mammals in the wild for the purpose of scientific research, the taking authorized under the permit involves such Level B harassment of marine mammals or marine mammal stocks, and they comply with the terms and conditions of that permit.

(2) Except as provided under paragraph (a)(1)(ii) of this section, no taking, including harassment, of marine mammals listed as threatened or endangered under the ESA is authorized under the General Authorization. Marine mammals listed as endangered or threatened under the ESA may be taken for purposes of scientific research only after issuance of a permit for such activities pursuant to the ESA.

(3) The following types of research activities will likely qualify for inclusion under the General Authorization:

Photo-identification studies, behavioral observations, and vessel and aerial population surveys (except aerial surveys over pinniped rookeries at altitudes of less than 1,000 ft).

(b) Letter of intent. Except as provided under paragraph (a)(1)(ii) of this section, any person intending to take marine mammals in the wild by Level B harassment for purposes of bona fide scientific research under the General Authorization must submit, at least 60 days before commencement of such research, a letter of intent by certified return/receipt mail to the Chief, Permits Division, F/PR1, Office of Protected Resources, NMFS, 1335 East-West Highway, Silver Spring, MD 20910-3226.

(1) The letter of intent must be submitted by the principal investigator (who shall be deemed the applicant). For purposes of this section, the principal investigator is the individual who is responsible for the overall research project, or the institution, governmental entity, or corporation responsible for supervision of the principal investigator.

(2) The letter of intent must include the following information:

(i) The name, address, telephone number, qualifications and experience of the applicant and any co-investigator(s) to be conducting the proposed research, and a curriculum vitae for each, including a list of publications by each such investigator relevant to the objectives, methodology, or other aspects of the proposed research;

(ii) The species or stocks of marine mammals (common and scientific names) that are the subject of the scientific research and any other species or stock of marine mammals that may be harassed during the conduct of the research;

(iii) The geographic location(s) in which the research is to be conducted, e.g., geographic name or lat./long.;

(iv) The period(s) of time over which the research will be conducted (up to five years), including the field season(s) for the research, if applicable;

(v) The purpose of the research, including a description of how the proposed research qualifies as bona fide research as defined in § 216.3; and

(vi) The methods to be used to conduct the research.

(3) The letter of intent must be signed, dated, and certified by the applicant as follows:

In accordance with section 104(c)(3)(C) of the Marine Mammal Protection Act of 1972, as amended (16 U.S.C. 1361 et seq.) and implementing regulations (50 CFR part 216), I hereby notify the National Marine Fisheries Service of my intent to conduct research involving only Level B harassment on marine mammals in the wild, and request confirmation that the General Authorization for Level B Harassment for Scientific Research applies to the proposed research as described herein. I certify that the information in this letter of intent is complete, true, and correct to the best of my knowledge and belief, and I understand that any false statement may subject me to the criminal penalties of 18 U.S.C. 1001, or penalties under the MMPA and implementing regulations. I acknowledge and accept that authority to conduct scientific research on marine mammals in the wild under the General Authorization is a limited conditional authority restricted to Level B harassment only, and that any other take of marine mammals, including the conduct of any activity that has the potential to injure marine mammals (i.e., Level A harassment), may subject me to penalties under the MMPA and implementing regulations.

(c) Confirmation that the General Authorization applies or notification of permit requirement.

(1) Not later than 30 days after receipt of a letter of intent as described in paragraph (b) of this section, the Chief, Permits Division, NMFS will issue a letter to the applicant either:

(i) Confirming that the General Authorization applies to the proposed scientific research as described in the letter of intent;

(ii) Notifying the applicant that all or part of the research described in the letter of intent is likely to result in a taking of a marine mammal in the wild involving other than Level B harassment and, as a result, cannot be conducted under the General Authorization, and that a scientific research permit is required to conduct all or part of the subject research; or

(iii) Notifying the applicant that the letter of intent fails to provide sufficient information and providing a description of the deficiencies, or notifying the applicant that the proposed research as described in the letter of intent is not bona fide research as defined in § 216.3.

(2) A copy of each letter of intent and letter confirming that the General Authorization applies or notifying the applicant that it does not apply will be forwarded to the Marine Mammal Commission.

(3) Periodically, NMFS will publish a summary document in the FEDERAL REGISTER notifying the public of letters of confirmation issued.

(d) Terms and conditions. Persons issued letters of confirmation in accordance with paragraph (c) of this section are responsible for complying with the following terms and conditions:

(1) Activities are limited to those conducted for the purposes, by the means, in the locations, and during the periods of time described in the letter of intent and acknowledged as authorized under the General

Authorization in the confirmation letter sent pursuant to paragraph (c) of this section;

(2) Annual reports of activities conducted under the General Authorization must be submitted to the Chief, Permits Division (address listed in paragraph (b) of this section) within 90 days of completion of the last field season(s) during the calendar year or, if the research is not conducted during a defined field season, no later than 90 days after the anniversary date of the letter of confirmation issued under paragraph (c) of this section. Annual reports must include:

(i) A summary of research activities conducted;

(ii) Identification of the species and number of each species taken by Level B harassment;

(iii) An evaluation of the progress made in meeting the objectives of the research as described in the letter of intent; and

(iv) Any incidental scientific, educational, or commercial uses of photographs, videotape, and film obtained as a result of or incidental to the research and if so, names of all photographers.

(3) Authorization to conduct research under the General Authorization is for the period(s) of time identified in the letter of intent or for a period of 5 years from the date of the letter of confirmation issued under paragraph (c) of this section, whichever is less, unless extended by the Director or modified, suspended, or revoked in accordance with paragraph (e) of this section;

(4) Activities conducted under the General Authorization may only be conducted under the on-site supervision of the principal investigator or co-investigator(s) named in the letter of intent. All personnel involved in the conduct of activities under the General Authorization must perform a function directly supportive of and necessary for the research being conducted, or be one of a reasonable number of support personnel included for the purpose of training or as back-up personnel;

(5) The principal investigator must notify the appropriate Regional Director, NMFS, in writing at least 2 weeks before initiation of on-site activities. The Regional Director shall consider this information in efforts to coordinate field research activities to minimize adverse impacts on marine mammals in the wild. The principal investigator must cooperate with coordination efforts by the Regional Director in this regard;

(6) If research activities result in a taking which exceeds Level B harassment, the applicant shall:

(i) Report the taking within 12 hours to the Director, Office of Protected Resources, or his designee as set forth in the letter authorizing research; and

(ii) Temporarily discontinue for 72 hours all field research activities that resulted in the taking. During this time period, the applicant shall consult with NMFS as to the circumstances surrounding the taking and any precautions necessary to prevent future taking, and may agree to amend the research protocol, as deemed necessary by NMFS.

(7) NMFS may review scientific research conducted pursuant to the General Authorization. If requested by NMFS, the applicant must cooperate with any such review and shall:

(i) Allow any employee of NOAA or any other person designated by the Director, Office of Protected Resources to observe research activities; and

(ii) Provide any documents or other information relating to the scientific research;

(8) Any photographs, videotape, or film obtained during the conduct of research under the General Authorization must be identified by a statement that refers to the General Authorization or ESA permit number, and includes the file number provided by NMFS in the confirmation letter, the name of the photographer, and the date the image was taken. This statement must accompany the image(s) in all subsequent uses or sales. The annual report must note incidental scientific, educational, or commercial uses of the images, and if there are any such uses, the names of all photographers; and

(9) Persons conducting scientific research under authority of the General Authorization may not transfer or assign any authority granted thereunder to any other person.

(e) Suspension, revocation, or modification.

(1) NMFS may suspend, revoke, or modify the authority to conduct scientific research under the General Authorization if:

(i) The letter of intent included false information or statements of a material nature;

(ii) The research does not constitute bona fide scientific research;

(iii) Research activities result in takings of marine mammals other than by Level B harassment;

(iv) Research activities differ from those described in the letter of intent submitted by the applicant and letter of confirmation issued by NMFS; or

(v) The applicant violates any term or condition set forth in this section.

(2) Any suspension, revocation, or modification is subject to the requirements of 15 CFR part 904.

SUBPART I—GENERAL REGULATIONS GOVERNING SMALL TAKES OF MARINE MAMMALS INCIDENTAL TO SPECIFIED ACTIVITIES

§ 216.101 Purpose.

The regulations in this subpart implement section 101(a)(5) (A) through (D) of the Marine Mammal Protection Act of 1972, as amended, 16 U.S.C. 1371(a)(5), which provides a mechanism for allowing, upon request, the incidental, but not intentional, taking of small numbers of marine mammals by U.S. citizens who engage in a specified activity (other than commercial fishing) within a specified geographic region.

§ 216.102 Scope.

The taking of small numbers of marine mammals under section 101(a)(5) (A) through (D) of the Marine Mammal Protection Act may be allowed only if the National Marine Fisheries Service:

(a) Finds, based on the best scientific evidence available, that the total taking by the specified activity during the specified time period will have a negligible impact on species or stock of marine mammal(s) and will not have an unmitigable adverse impact on the availability of those species or stocks of marine mammals intended for subsistence uses;

(b) Prescribes either regulations under § 216.106, or requirements and conditions contained within an incidental harassment authorization issued under § 216.107, setting forth permissible methods of taking and other means of effecting the least practicable adverse impact on the species or stock of marine mammal and its habitat and on the availability of the species or stock of marine mammal for subsistence uses, paying particular attention to rookeries, mating grounds, and areas of similar significance; and

(c) Prescribes either regulations or requirements and conditions contained within an incidental harassment authorization, as appropriate, pertaining to the monitoring and reporting of such taking. The specific regulations governing certain specified activities are contained in subsequent subparts of this part.

§ 216.103 Definitions.

In addition to definitions contained in the MMPA, and in § 216.3, and unless the context otherwise requires, in subsequent subparts to this part:

Incidental harassment, incidental taking and incidental, but not intentional, taking all mean an accidental taking. This does not mean that the taking is unexpected, but rather it includes those takings that are infrequent, unavoidable or accidental. (A complete definition of ''take'' is contained in § 216.3).

Negligible impact is an impact resulting from the specified activity that cannot be reasonably expected to, and is not reasonably likely to, adversely affect the species or stock through effects on annual rates of recruitment or survival.

Small numbers means a portion of a marine mammal species or stock whose taking would have a negligible impact on that species or stock.

§ 216.104 Submission of requests.

(a) In order for the National Marine Fisheries Service to consider authorizing the taking by U.S. citizens of small numbers of marine mammals incidental to a specified activity (other than commercial fishing), or to make a finding that an incidental take is unlikely to occur, a written request must be submitted to the Assistant Administrator. All requests must include the following information for their activity:

(1) A detailed description of the specific activity or class of activities that can be expected to result in incidental taking of marine mammals;

(2) The date(s) and duration of such activity and the specific geographical region where it will occur;

(3) The species and numbers of marine mammals likely to be found within the activity area;

(4) A description of the status, distribution, and seasonal distribution (when applicable) of the affected species or stocks of marine mammals likely to be affected by such activities;

(5) The type of incidental taking authorization that is being requested (i.e., takes by harassment only; takes by harassment, injury and/or death) and the method of incidental taking;

(6) By age, sex, and reproductive condition (if possible), the number of marine mammals (by species) that may be taken by each type of taking identified in paragraph (a)(5) of this section, and the number of times such takings by each type of taking are likely to occur;

(7) The anticipated impact of the activity upon the species or stock of marine mammal;

(8) The anticipated impact of the activity on the availability of the species or stocks of marine mammals for subsistence uses;

(9) The anticipated impact of the activity upon the habitat of the marine mammal populations, and the likelihood of restoration of the affected habitat;

(10) The anticipated impact of the loss or modification of the habitat on the marine mammal populations involved;

(11) The availability and feasibility (economic and technological) of equipment, methods, and manner of conducting such activity or other means of effecting the least practicable adverse impact upon the affected species or stocks, their habitat, and on their availability for subsistence

uses, paying particular attention to rookeries, mating grounds, and areas of similar significance;

(12) Where the proposed activity would take place in or near a traditional Arctic subsistence hunting area and/or may affect the availability of a species or stock of marine mammal for Arctic subsistence uses, the applicant must submit either a plan of cooperation or information that identifies what measures have been taken and/or will be taken to minimize any adverse effects on the availability of marine mammals for subsistence uses. A plan must include the following:

(i) A statement that the applicant has notified and provided the affected subsistence community with a draft plan of cooperation;

(ii) A schedule for meeting with the affected subsistence communities to discuss proposed activities and to resolve potential conflicts regarding any aspects of either the operation or the plan of cooperation;

(iii) A description of what measures the applicant has taken and/or will take to ensure that proposed activities will not interfere with subsistence whaling or sealing; and

(iv) What plans the applicant has to continue to meet with the affected communities, both prior to and while conducting the activity, to resolve conflicts and to notify the communities of any changes in the operation;

(13) The suggested means of accomplishing the necessary monitoring and reporting that will result in increased knowledge of the species, the level of taking or impacts on populations of marine mammals that are expected to be present while conducting activities and suggested means of minimizing burdens by coordinating such reporting requirements with other schemes already applicable to persons conducting such activity. Monitoring plans should include a description of the survey techniques that would be used to determine the movement and activity of marine mammals near the activity site(s) including migration and other habitat uses, such as feeding. Guidelines for developing a site-specific monitoring plan may be obtained by writing to the Director, Office of Protected Resources; and

(14) Suggested means of learning of, encouraging, and coordinating research opportunities, plans, and activities relating to reducing such incidental taking and evaluating its effects.

(b)

(1) The Assistant Administrator shall determine the adequacy and completeness of a request and, if determined to be adequate and complete, will begin the public review process by publishing in the FEDERAL REGISTER either:

(i) A proposed incidental harassment authorization; or

(ii) A notice of receipt of a request for the implementation or re-implementation of regulations governing the incidental taking.

(2) Through notice in the FEDERAL REGISTER, newspapers of general circulation, and appropriate electronic media in the coastal areas that may be affected by such activity, NMFS will invite information, suggestions, and comments for a period not to exceed 30 days from the date of publication in the FEDERAL REGISTER. All information and suggestions will be considered by the National Marine Fisheries Service in developing, if appropriate, the most effective regulations governing the issuance of letters of authorization or conditions governing the issuance of an incidental harassment authorization.

(3) Applications that are determined to be incomplete or inappropriate for the type of taking requested, will be returned to the applicant with an explanation of why the application is being returned.

(c) The Assistant Administrator shall evaluate each request to determine, based upon the best available scientific evidence, whether the taking by the specified activity within the specified geographic region will have a negligible impact on the species or stock and, where appropriate, will not have an unmitigable adverse impact on the availability of such species or stock for subsistence uses. If the Assistant Administrator finds that the mitigating measures would render the impact of the specified activity negligible when it would not otherwise satisfy that requirement, the Assistant Administrator may make a finding of negligible impact subject to such mitigating measures being successfully implemented. Any preliminary findings of "negligible impact" and "no unmitigable adverse impact" shall be proposed for public comment along with either the proposed incidental harassment authorization or the proposed regulations for the specific activity.

(d) If, subsequent to the public review period, the Assistant Administrator finds that the taking by the specified activity would have more than a negligible impact on the species or stock of marine mammal or would have an unmitigable adverse impact on the availability of such species or stock for subsistence uses, the Assistant Administrator shall publish in the FEDERAL REGISTER the negative finding along with the basis for denying the request.

§ 216.105 Specific regulations.

(a) For all petitions for regulations under this paragraph, applicants must provide the information requested in § 216.104(a) on their activity as a whole, which includes, but is not necessarily limited to, an assessment of total impacts by all persons conducting the activity.

(b) For allowed activities that may result in incidental takings of small numbers of marine mammals by harassment, serious injury, death or a combination thereof, specific regulations shall be established for each allowed activity that set forth:

(1) Permissible methods of taking;

(2) Means of effecting the least practicable adverse impact on the species and its habitat and on the availability of the species for subsistence uses; and

(3) Requirements for monitoring and reporting, including requirements for the independent peer-review of proposed monitoring plans where the proposed activity may affect the availability of a species or stock for taking for subsistence uses.

(c) Regulations will be established based on the best available information. As new information is developed, through monitoring, reporting, or research, the regulations may be modified, in whole or in part, after notice and opportunity for public review.

§ 216.106 Letter of Authorization.

(a) A Letter of Authorization, which may be issued only to U.S. citizens, is required to conduct activities pursuant to any regulations established under § 216.105. Requests for Letters of Authorization shall be submitted to the Director, Office of Protected Resources. The information to be submitted in a request for an authorization will be specified in the appropriate sub-part to this part or may be obtained by writing to the above named person.

(b) Issuance of a Letter of Authorization will be based on a determination that the level of taking will be consistent with the findings made for the total taking allowable under the specific regulations.

(c) Letters of Authorization will specify the period of validity and any additional terms and conditions appropriate for the specific request.

(d) Notice of issuance of all Letters of Authorization will be published in the FEDERAL REGISTER within 30 days of issuance.

(e) Letters of Authorization shall be withdrawn or suspended, either on an individual or class basis, as appropriate, if, after notice and opportunity for public comment, the Assistant Administrator determines that:

(1) The regulations prescribed are not being substantially complied with; or

(2) The taking allowed is having, or may have, more than a negligible impact on the species or stock or, where relevant, an unmitigable adverse impact on the availability of the species or stock for subsistence uses.

(f) The requirement for notice and opportunity for public review in § 216.106(e) shall not apply if the Assistant Administrator determines that an emergency exists that poses a significant risk to the well-being of the species or stocks of marine mammals concerned.

(g) A violation of any of the terms and conditions of a Letter of Authorization or of the specific regulations shall subject the Holder and/or any individual who is operating under the authority of the Holder's Letter of Authorization to penalties provided in the MMPA.

§ 216.107 Incidental harassment authorization for Arctic waters.

(a) Except for activities that have the potential to result in serious injury or mortality, which must be authorized under § 216.105, incidental harassment authorizations may be issued, following a 30-day public review period, to allowed activities that may result in only the incidental harassment of a small number of marine mammals. Each such incidental harassment authorization shall set forth:

(1) Permissible methods of taking by harassment;

(2) Means of effecting the least practicable adverse impact on the species, its habitat, and on the availability of the species for subsistence uses; and

(3) Requirements for monitoring and reporting, including requirements for the independent peer-review of proposed monitoring plans where the proposed activity may affect the availability of a species or stock for taking for subsistence uses.

(b) Issuance of an incidental harassment authorization will be based on a determination that the number of marine mammals taken by harassment will be small, will have a negligible impact on the species or stock of marine mammal(s), and will not have an unmitigable adverse impact on the availability of species or stocks for taking for subsistence uses.

(c) An incidental harassment authorization will be either issued or denied within 45 days after the close of the public review period.

(d) Notice of issuance or denial of an incidental harassment authorization will be published in the FEDERAL REGISTER within 30 days of issuance of a determination.

(e) Incidental harassment authorizations will be valid for a period of time not to exceed 1 year but may be renewed for additional periods of time not to exceed 1 year for each reauthorization.

(f) An incidental harassment authorization shall be modified, withdrawn, or suspended if, after notice and opportunity for public comment, the Assistant Administrator determines that:

(1) The conditions and requirements prescribed in the authorization are not being substantially complied with; or

(2) The authorized taking, either individually or in combination with other authorizations, is having, or may have, more than a negligible impact on the species or stock or, where relevant, an unmitigable adverse impact on the availability of the species or stock for subsistence uses.

(g) The requirement for notice and opportunity for public review in paragraph (f) of this section shall not apply if the Assistant Administrator determines that an emergency exists that poses a significant risk to the well-being of the species or stocks of marine mammals concerned.

(h) A violation of any of the terms and conditions of an incidental harassment authorization shall subject the holder and/or any individual who is operating under the authority of the holder's incidental harassment authorization to penalties provided in the MMPA.

§ 216.108 Requirements for monitoring and reporting under incidental harassment authorizations for Arctic waters.

(a) Holders of an incidental harassment authorization in Arctic waters and their employees, agents, and designees must cooperate with the National Marine Fisheries Service and other designated Federal, state, or local agencies to monitor the impacts of their activity on marine mammals. Unless stated otherwise within an incidental harassment authorization, the holder of an incidental harassment authorization effective in Arctic waters must notify the Alaska Regional Director, National Marine Fisheries Service, of any activities that may involve a take by incidental harassment in Arctic waters at least 14 calendar days prior to commencement of the activity.

(b) Holders of incidental harassment authorizations effective in Arctic waters may be required by their authorization to designate at least one qualified biological observer or another appropriately experienced individual to observe and record the effects of activities on marine mammals. The number of observers required for monitoring the impact of the activity on marine mammals will be specified in the incidental harassment authorization. If observers are required as a condition of the authorization, the observer(s) must be approved in advance by the National Marine Fisheries Service.

(c) The monitoring program must, if appropriate, document the effects (including acoustical) on marine mammals and document or estimate the actual level of take. The requirements for monitoring plans, as specified in the incidental harassment authorization, may vary depending on the activity, the location, and the time.

(d) Where the proposed activity may affect the availability of a species or stock of marine mammal for taking for subsistence purposes, proposed monitoring plans or other research proposals must be independently peer-reviewed prior to issuance of an incidental harassment authorization under this subpart. In order to complete the peer-review process within the time frames mandated by the MMPA for an incidental harassment authorization, a proposed monitoring plan submitted under this paragraph must be submitted to the Assistant Administrator no later than the date of submission of the application for an incidental harassment authorization. Upon receipt of a complete monitoring plan, and at its discretion, the National Marine Fisheries Service will either submit the plan to members of a peer review panel for review or within 60 days of receipt of the proposed monitoring plan, schedule a workshop to review the plan. The applicant must submit a final monitoring plan to the Assistant Administrator prior to the issuance of an incidental harassment authorization.

(e) At its discretion, the National Marine Fisheries Service may place an observer aboard vessels, platforms, aircraft, etc., to monitor the impact of activities on marine mammals.

(f)

(1) As specified in the incidental harassment authorization, the holder of an incidental harassment authorization for Arctic waters must submit reports to the Assistant Administrator within 90 days of completion of any individual components of the activity (if any), within 90 days of completion of the activity, but no later than 120 days prior to expiration of the incidental harassment authorization, whichever is earlier. This report must include the following information:

(i) Dates and type(s) of activity;

(ii) Dates and location(s) of any activities related to monitoring the effects on marine mammals; and

(iii) Results of the monitoring activities, including an estimate of the actual level and type of take, species name and numbers of each species observed, direction of movement of species, and any observed changes or modifications in behavior.

(2) Monitoring reports will be reviewed by the Assistant Administrator and, if determined to be incomplete or inaccurate, will be returned to the holder of the authorization with an explanation of why the report is being returned. If the authorization holder disagrees with the findings of the Assistant Administrator, the holder may request an independent peer review of the report. Failure to submit a complete and accurate report may result in a delay in processing future authorization requests.

(g) Results of any behavioral, feeding, or population studies, that are conducted supplemental to the monitoring program, should be made available to the National Marine Fisheries Service before applying for an incidental harassment authorization for the following year.

OSHA REGULATIONS (STANDARDS - 29 CFR)[2]
OCCUPATIONAL NOISE EXPOSURE. - 1910.95
(SELECTED PORTIONS)

(a) Protection against the effects of noise exposure shall be provided when the sound levels exceed those shown in Table G-16 when measured on the A scale of a standard sound level meter at slow response. When noise levels are determined by octave band analysis, the equivalent A-weighted sound level may be determined as follows (Figure G-9):

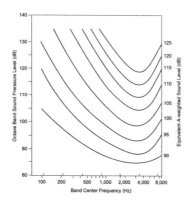

FIGURE G-9 - Equivalent A-Weighted Sound Level

─────────────────

[2] www.osha-slc.gov/OshStd_data/1910_0095.html, accessed 9/6/99.

Equivalent sound level contours. Octave band sound pressure levels may be converted to the equivalent A-weighted sound level by plotting them on this graph and noting the A-weighted sound level corresponding to the point of highest penetration into the sound level contours. This equivalent A-weighted sound level, which may differ from the actual A-weighted sound level of the noise, is used to determine exposure limits from Table G-16.

(b)

(1) When employees are subjected to sound exceeding those listed in Table G-16, feasible administrative or engineering controls shall be utilized. If such controls fail to reduce sound levels within the levels of Table G-16, personal protective equipment shall be provided and used to reduce sound levels within the levels of the table.

(2) If the variations in noise level involve maxima at intervals of 1 second or less, it is to be considered continuous.

TABLE G-16 - PERMISSIBLE NOISE EXPOSURES (1)

Duration per day, hours	Sound level dBA slow response
8	90
6	92
4	95
3	97
2	100
$1^1/_2$	102
1	105
$^1/_2$	110
$^1/_4$ or less	115

Footnote(1) When the daily noise exposure is composed of two or more periods of noise exposure of different levels, their combined effect should be considered, rather than the individual effect of each. If the sum of the following fractions: $C(1)/T(1) + C(2)/T(2) \ C(n)/T(n)$ exceeds unity, then, the mixed exposure should be considered to exceed the limit value. Cn indicates the total time of exposure at a specified noise level, and Tn indicates the total time of exposure permitted at that level. Exposure to impulsive or impact noise should not exceed 140 dB peak sound pressure level.

(10) "Standard threshold shift."

(i) As used in this section, a standard threshold shift is a change in hearing threshold relative to the baseline audiogram of an average of 10 dB or more at 2000, 3000, and 4000 Hz in either ear.

(ii) In determining whether a standard threshold shift has occurred, allowance may be made for the contribution of aging (presbycusis) to the change in hearing level by correcting the annual audiogram according to the procedure described in Appendix F: "Calculation and Application of Age Correction to Audiograms."

E — Glossary of Acronyms

ABR	auditory brainstem response
AEP	auditory evoked potential
ATOC	Acoustic Thermometry of Ocean Climate experiment
CITES	Convention on International Trade in Endangered Species of Wild Fauna and Flora
DARPA	Defense Advanced Research Projects Agency
EIS	environmental impact statement
GCM	general circulation model
HIFT	Heard Island Feasibility Test
IHA	incidental harassment authorization
IUSS	Integrated Undersea Surveillance System
LFA	low-frequency active (sonar)
MMPA	Marine Mammal Protection Act
MMRP	Marine Mammal Research Program
MMS	Minerals Management Service
MTTS	masked temporary threshold shift

NIH	National Institutes of Health
NMFS	National Marine Fisheries Service
NOAA	National Oceanic and Atmospheric Administration
NRC	National Research Council
NRDC	Natural Resources Defense Council
NSF	National Science Foundation
ONR	Office of Naval Research
PBR	potential biological removal
PTS	permanent threshold shift
RFP	Request for Proposal
SOFAR	SOund Fixing And Ranging
SOSUS	U.S. Navy's SOund SUrveillance System
SPL	sound pressure level
SWAT	Stranded Whale Auditory Test (team)
TTS	temporary threshold shift

F

Species Mentioned in This Report

Common Name	Scientific Name
Cetaceans	
Odontocetes (toothed whales)	
Sperm whale	*Physeter macrocephalus*
Pygmy sperm whale	*Kogia breviceps*
Pilot whale	*Globicephala sp.*
White whale	*Delphinapterus leucas*
Bottlenose dolphin	*Tursiops truncatus*
Killer whale	*Orcinus orca*
False killer whale	*Pseudorca crassidens*
Risso's dolphin	*Grampus griseus*
Cuvier's beaked whale	*Ziphius cavirostris*
Mysticetes (baleen whales)	
Northern right whale	*Eubalaena glacialis*
Humpback whale	*Megaptera novaeangliae*
Gray whale	*Eschrichtius robustus*
Fin/Finback whale	*Balaenoptera physalus*
Blue whale	*Balaenoptera musculus*
Pygmy blue whale	*Balaenoptera musculus brevicauda*

Pinnipeds
    Phocids ("true" eared seals)
        Harbor seal   *Phoca vitulina*
        Northern elephant seal   *Mirounga angustirostris*
    Otariids (fur seals and sea lions)
        California sea lion   *Zalophus californianus*

Sea Turtles
    Green turtle   *Chelonia mydas*
    Hawksbill turtle   *Eretmochelys imbricata*
    Leatherback turtle   *Dermochelys coriacea*
    Olive Ridley turtle   *Lepidochelys olivacea*
Fish
    Sheepshead minnow   *Cyprinodon variegatus*
    Killifish   *Fundulus similis*

Crustaceans
    Shrimp   *Crangon crangon*

DATE DUE